The Evolution of
Human Sexual Privacy

The Evolution of Human Sexual Privacy

An Objective Study of a Subjective Realm

Andrew Haywein

Outskirts Press, Inc.
Denver, Colorado

Outskirts Press, Inc.
http://www.outskirtspress.com

ISBN: 978-1-4327-7372-4

Library of Congress Control Number: 2011914828

Outskirts Press and the "OP" logo are trademarks belonging to Outskirts Press, Inc.

PRINTED IN THE UNITED STATES OF AMERICA

For humankind's vast future,

a glimpse into a formative period long past.

Writing only brings out comparatively superficial experiences. Man has had it for a relatively short time – shall we say about four thousand years? ... Now for ages before that you had immense quantities of human experience accumulating in men's bodies. The body itself was, and is, an immense experience; the sheer harmony of its properly functioning organs gives us a flood of unconscious enjoyment. It is quite inarticulate, and doesn't need to be articulate. But in bulk, and perhaps in significance, it far outweighs the scope of the written word.

- A. N. Whitehead

It is very often the case that the behavior of an animal is inexplicable in the context of the present. Only by studying the past does it become clear.

- J. Napier and T. Napier

The introduction to this felicity is in a tender and private relation of one to one, which is the enchantment of human life.

- R. W. Emerson

CONTENTS

PREFACE

The intense privacy that surrounds the individual sexual experience tends to preclude the transmission of useful knowledge of this realm from one generation to the next. The human sexual situation is quite unique, for the very essence of culture has been the transmission of helpful knowledge from generation to generation. How has this come about? What are its consequences? Has the emergence of sexual privacy made the solving of sexual problems more difficult? Does the insulation of privacy tend to limit sexual fulfillment? Beyond culture, how is generic human sexual fulfillment to be envisioned? The present study explores these questions.

Knowledge of agriculture, health, commerce, law, engineering, history, or any field of human endeavor other than the sexual is taught in schools or, in one way or another, is transmitted by culture. Many responsible parents, wishing to prevent mistakes, would like to pass on basic information relating to sex to their children, but find this task so emotionally difficult as to prove beyond their capacities. Even the best of friends rarely discuss matters relating to sexual intimacy. Yet, sexual reproduction is of great importance to the individual and is essential for the continuance of the species. How could this curious cultural bind have come about, and what might be its answer?

We immediately note that it is the very nature of the human sexual relationship to be tenaciously private. Every person wants it to be so, and, indeed, would be much troubled if it were not. In this respect, humans have advanced well beyond the level of the great apes, who engage in sex openly and casually. At the human level privacy has become a precondition, both for the sexual experience itself and for its aftermath. As individuals are maturing in early adulthood, each looks forward to exploring and learning on one's own in this most intimate of life's realms. What happens to us individually in this sphere is our own business. Strong inner feelings affirm that this is quite as it should be. It would be wonderful if only life in this realm were simple, if problems did not so commonly arise.

Human sexual privacy, however, by its very nature involves consequences and invites problems. The major consequence appears to relate to the fact that every human lifetime involves sexual learning. Every human personality becomes defined, in part, by its sexual history. In the life course progression of each person, sexual learning inevitably encounters two major problems.

First, when intercourse commences in early adulthood the sexual need profiles of the human genders show a marked disparity. While the human male at this stage experiences the highest sexual intensity of his lifetime, the woman finds herself maximally inhibited. Years of shared experience are often needed for full sexual harmony to develop, which is not always fully attained. It is surprising that the human sexual career should begin with such a handicap. As they are entering upon their new sexual world, couples in some cultures receive advice from older members, but all such tends to be scant in particulars, and most seem to give no instruction at all. Our own culture makes little regular attempt to advise young persons as they enter into this adult realm.

Second, by its nature sexual learning is deeply involved with emotional learning. In the course of sexual learning offenses and mistakes commonly occur, perhaps unavoidably, causing emotional injury to others as well as to the self. Though enshrined in privacy, lesser sexually relating problems are usually able to be managed rationally or to become emotionally outgrown. Larger sexual problems, however, arising from more involved or traumatic experiences, prove all the more difficult to resolve because of their inherently guarded state. When they persist, unresolved problems show significant capacity to influence the still developing personality, at times entirely redirecting it. At any age, however, unresolved sexual problems tend to act restrictively upon the persons involved.

We are witnessing today a veritable deluge of sexual information which at first might be thought to be filling this cultural void. However, the curious feature of this knowledge is that sexual information alone and even sexual experience alone do not finally satisfy. Human sexual relationships are notoriously complicated. That which is needed involves a complementary understanding of the emotional aspects of sexual relationships and a sense of gender related interpersonal dynamics. Without including these dimensions as working knowledge, sexual information alone proves frequently unable to avoid trouble.

Attempting to correct sexually related and emotional problems in modern culture, the psychiatric disciplines have quite prospered. Their help, however, often tends to be limited. Frequently personal damage has been experienced for which psychiatric correction proves to be but partial. What seems to be needed particularly is a broad cultural perspective so improving understanding that sexual problems may become significantly reduced or, even better, prevented.

Although the problem of human sexual nature is broad beyond ordinary measure, many experts have contributed major gains toward its clarification. Important strides have been made. In many ways we are far more advanced today in our sexually relating knowledge than any previous generation. We now understand the basic physiology of sexual experience, for instance, quite well. Mainly for want of comprehensive perspective, however, no approach has escaped criticism. Each major study has reflected the limitations inherent in the perspective and variables of the particular approach. The remarkable pioneering work of Kinsey and his associates has been criticized for omitting the emotional dimension. Masters and Johnson, their successors, using terms such as *outlets* and *plateaus*, have been described as treating this highly personal subject impersonally, as would plumbers or carpenters. Clinical psychiatrists study human sexuality intensely, but with an emphasis upon overcoming pathology rather than upon finding fulfillment. Comparative primate studies have revealed wide variation in less evolutionally advanced sexual behaviors, with none fully qualifying as a satisfactory human model. Cultural anthropologists have revealed an enormously wide range of cross cultural diversity in human sexual practices but are criticized for offering no well defined central modality. Working today with ever more sophisticated technological methods and increasing confidence, physical anthropologists and ecologists are conscientiously charting human origins and their environmental contexts with a promisingly broad scope and an unprecedented scientific accuracy. However, even here, large gaps of understanding remain. The driven sexuality of a great ape female in estrus, our closest evolutionary kin, remains well removed from that of a normal modern woman. Clearly, the defining reproductive advances between the level of the great apes and modern humans have been major but remain largely uncharted.

It is becoming clear that the sexuality of a human today represents an enormous inheritance, the summation of countless lifetimes over numberless generations of men and women who have loved, worked for each other and together raised their families. Their primary record, however poorly discernable, lies in human nature today. For attaining a comprehensive perspective upon this, the approach which would seem the most promising would need to be one holistic and developmental. As the geneticist Dobzhansky aptly observed, it is only in the light of evolution that advanced biological function makes sense.

Accordingly, the present study approaches human sexuality through an attempt to further understand its evolutional development. This attempts to envision components of an overall framework for human reproductive processes as they have evolved since departure from a common ancestor with the great apes. Fortunately, based upon evidence from the fossil record, paleontologists currently present us with a fairly clear account of many of the salient features of this seven million-year interval. Not only do stones and bones leave traces, however, but also, in admittedly less apparent ways, so do the soft tissue processes of existing humans. Late evolutional developments in the human nervous system, which tripled in its size during this temporal interval, in the more complex modern endocrine system, in the unique human skin with its hairlessness, and in our distinctly human genitals present clear evolutional developments awaiting fuller interpretation.

The lifetime of the author, long an amateur naturalist and a generalist in life science, fortunately lent itself to such a study. A medical career with sustained interests in endocrinology and gynecology commenced with ten years of medical practice, from 1958 to 1969, managing a rural African bush hospital. The indigenous culture at the time was not far removed from Stone Age ways. Local members of primate species, particularly baboons, were numerous. From the edge of a nearby tropical rainforest the large Simango monkeys made frequent forays onto the ground, stealing vegetables and raiding fruit trees. At night the small lemuroid bush babies, with eyes reflecting like headlights, descended to raid banana trees. Although the indigenous tribe was Bantu, consisting of agriculturalists and pastoralists, the world of the Bushmen, a remaining culture of hunters and gatherers to the south, seemed never far away.

Memories of these years provide a lasting sense of a far simpler way of life, one not too far removed from that within which human nature received its basic definition. Centering about hunting and gathering, the primary human culture remained remarkably stable and uniform for millennia upon millennia, up to the historically recent introduction of agriculture. Although the real reason for a particular event in such a culture might not always be that given, the underlying cultural dynamics of the happenings of everyday life consistently made good sense. Time and again when confronted by problems of modern life the author has found himself reflecting back upon this earlier human context in which all made sense. This has happened not only for some of life's larger problems but also in reflecting upon everyday events. When a baseball pitcher on the mound throws his ball with such consummate skill, it seems but a modern version of a hunter hurling his spear at an impala, one perhaps finally worn down after long tracking. When a woman walks, the sway of her hips makes possible the level carrying of a load on her head, perhaps a bundle of firewood or a pitcher of water, household tasks which are no longer daily duties. In this setting, human sexuality was encountered not as fragmented behavior but as an integral component of the daily lives of the culture. Yet, in truth, behind all such mental meanderings has remained the amateur naturalist, from early boyhood on wondering about this venture of human life.

It became apparent early in this study that an evolutional approach exploring the human past appeared capable of offering provisional answers to certain perplexing modern human reproductive problems. Woman's slow sexual arousal and her irregular response, for instance, considered to be a modern woman's problem, do not appear to reflect inherent pathology or evolutional failure of development, as is often suggested, but, rather, well programmed design. Requiring control and delay on the part of the male, such womanly features appear to have served to increase emotional bonding. Similarly, the distinct postcoital refractory period of the human male does not appear to represent unrefined evolutional crudity but, rather, also appears to be the product of evolving functional design. The ecologic imperative for the human male has been for him to obtain food for the family, particularly meat.

For this he must hunt. Were he not to have at least a short refractory period after sex, he might never leave home. Yet the overall functional design proves to be subtle. The refractory period is not long. Shortly, aroused by testosterone elaborated during the hunt, the male becomes desirous of returning home. The human female, experiencing no similar postcoital refractory state, can welcome him at any time.

As this study proceeded, it proved to be remarkable how the identification of two major landmarks led to the locating of so many others. The presumed estrogenic advance early in the woodland stage and the apparent androgenic advance as the hunter-gatherer way of life got underway led to the likely placement of many other actions of these hormones. Each of these major gender hormones appears to have introduced an entirely new cluster of elaborative functions which involve fields of concurrent change. If there were a rosetta stone to this study, it would have existed in the apparent evolutional placement of these two major hormonal advances. In its descriptions of processes of reproductive sensory summation, this study is also indebted to Wald for his appreciation of animate organization as a process of energy accumulation.

A perusal of the evolutional record suggests that there emerged a crucial turning point in human gender relations which served to inaugurate the distinctly human reproductive advance. This appears to have taken place when sexual relations changed from being public to private. This transition would trace the advent of privacy to the attainment of satisfactory ground safety, an achievement which initiated the erectine career. With the rise of privacy would have come the first use of dwellings, primarily caves and crude huts, for the transient home bases of a nomadic way of life.

The introduction of sexual privacy appears to have opened up an entirely new world of conscious meanings for human nature. This nascent realm appears to have centered first about shared responsibilities for the feeding and development of human young, now requiring longer years to mature. The new environmental changes which commenced at this time were requiring ever more complex human behaviors. Requiring greater developmental years to meet repeatedly more demanding environmental challenges, the young with their increased maturation required progressively greater familial nurturance under the guidance of both parents.

Centering about this focal evolutional change, the present study divides into three parts. The first examines the features in human nature which coalesce to produce a need for sexual privacy. Why should such needs, unique to humans, exist? The second part explores the evolutional features of the hominid state that appear to have led to privacy's emergence at human levels. Some of these initial developments continue as distinguishable components of human reproductive nature today. The third part explores various dimensions of the new world of human sexual experience which unfolded in subsequent evolution upon the basis of sexual privacy.

As this study progressed, the author came to realize how much its descriptions reflect life unfulfilled as life fulfilled. We appear not to be realizing in our time our fullest sexual potentialities. Fortunately, as well as with full levels of working reserves, human reproductive processes, having weathered much in evolution, can usually course quite adequately with less than optimal reserves. However, by a curious turn of thought while ruminating upon romance, the poet Dante came to mind. Had the relationship between Dante and his beloved Beatrice been one of fulfilled consummation, he would probably never have written more than a few words in her behalf, and possibly not even that. It requires to be appreciated that projects are undertaken as much in unresolved need as by satisfaction upon fulfillment. The fulfilled mind typically turns to other problems.

In such a light, this study essentially presents a position paper upon the interrelational status of man and woman in our day. The approach is not by review of current attitudes or practices but by attempting to discern the evolutional background of human gender natures. The value of any scientific or cross-cultural study lies in its potential for universal validity. The science of physiology and the hunter-gatherer culture, particularly its last half million years of remarkably well-defined stability, highly commend each of these fields for more inclusive use. The rural African Bantu culture of the mid-twentieth century which the author encountered cannot be equated with its earlier hunter-gatherer origins, but, on the other hand, it does qualify for providing a far more appropriate reference for understanding functional human gender meanings than any twenty-first-century culture or the state of civilization itself. It is hoped that such cross cultural perspectives may help to envision, in time, a representative overall framework of human soft tissue evolutional advances, well beyond the early interpretations of the present study.

A study such as the present, a pilot project, can claim no final or comprehensive validity. The subject is far too vast and the study itself far too incomplete. We of today are still in the process of trying to find out who we are. Rather, from a broad historical background that now has become sufficiently clear to suggest preliminary interpretation, this study primarily hopes to shed light upon some of our contemporary sexual problems, of which there are many. In advanced industrial societies today it is now estimated that more than half of all children are born into one parent households. It is the outstanding plight of the women in *Sex and the City* to be unable to find enduring love relationships. We need all the help from the past that we can find.

It must be appreciated for any comprehensive view that civilization has realized significant advancements in conjugal rewards. These, like civilization itself, are not yet universally present or, when present, always reliably sustained. Whereas a general easing of the burden of basic maintenance work tends to occur with the advance to civilized life, and, with it, a narrowing of gender roles, personality development tends to become more individuated. Civilization presents more specialized and elaborate ways of life. For couples, this elevates the weight of interpersonal relations into primacy compared with hunter-gatherer contexts. In civilized contexts lasting compatibility becomes more important, more difficult to find and more valued. This is the level of human interrelation that appears to need the greatest further engagement as humans face their future.

This study began in an attempt to explore the problem of the relevance of triply tiered human brain structure to the human sexual experience. It became readily apparent that the evolutional levels represented, respectively, by the thalamus, the limbic system, and the neocortex are distinguishable in human experience as, respectively, sexual need, emotional need, and personal need. Since the functional correlates of relationships at each level are different, the relative participation of each stratum in any given occasion of sexual experience largely portrays its general physiological and psychological expression. When existing alone, human sexual need is different from when it is alloyed with love, with interpersonal involvement, or with both. As this study draws to an initial summation, the author cannot but regard an appreciation of the triply tiered nature of human reproduction to be still the most central feature of sexual dynamics.

Perhaps the greatest surprise in this study was the realization that all of the varied aspects of human reproductive nature appear to direct toward one consistent end - namely, the improved biological investment in the young, in each unit progeny of a next generation. If the reader, desiring to save time, wishes to scan this study, its grand summation can be found in the last two paragraphs of Chapter XXI.

With roots so ancient, problems of sex and gender are neither easy to discuss nor always simple to understand. Further, as noted, the nature of human sexual privacy acts to insulate problems when they occur. In submitting this study, the author hopes that, beyond alleviating problems, an appreciation of their evolutional history may help not only to prevent them from occurring but, hopefully, to lead to more fulfilled personal lives.

As the reader will often sense, much of the material in this study, including that relating to traditional African culture, has been given to the author over the years in confidence. As a member of a very modest generation and desiring to render full respect to the subject of the study itself, the author, now well in his senior years, prefers to submit it under a pseudonym, requesting that the reader honor his wishes for privacy. Let the study, unfinished as it is, speak for itself.

Andrew Haywein

PART I: WHY SEXUAL PRIVACY?

WHY HUMAN SEXUAL PRIVACY?

*I*t is quite clear that an intense need for sexual privacy is inherent in human nature, but it remains equally unclear how it developed or what its full physiological role might be. No human experiences are as intimate or tend more to remain secret. No particulars are more likely to descend with a human being into the grave. We sense that somehow comfort, safety and cultural taboos are involved, but beyond the fact that intense privacy surrounds all human sexual experience explanation fails.

In many respects the situation is perplexing. Human reproduction is of a level of advanced biological complexity unmatched in any other vertebrate. All too often these complexities go awry. For every organism on earth reproduction is a quintessential life function. The fact that each generation is placed on its own means that the benefits of learning, a certain helpful, well-tested cultural sense of sexual direction, is lacking. As this study supports, whereas the simpler hunting and gathering ecology within which human nature developed during most of its evolutionary career provided a general background context within which sexual learning fitted smoothly, the contexts of civilized societies become removed from such corrective guidance. In this respect, humans in the far more developed and culturally diverse world of today, far removed from aboriginal ways, find themselves at vague disadvantage and in unknowing handicap. What was once clear to generation upon generation of our ancestors, a direct route of human sexual maturation, is no longer culturally reinforced or even clear.

In the conviction that a better understanding of human gender relationships as they developed in contexts both of changing environmental ecologies and of deepening sexual privacy should be a gain, the present study attempts to explore the comparative higher vertebrate world within which these reproductively relating human developments took place. In particular, it attempts to include soft tissue-tissue advances that leave no fossil traces but which, resulting in human physiology today, are so central in this long unfolding drama.

I

SEXUAL PRIVACY'S CURIOUS INNATE GUARDEDNESS

*O*bjectivity in relation to anything sexual cannot be counted to come with ease. It is hoped that a qualification claimed by the writer, that, as mentioned, he is now well along in his senior years, may help. In particular, however, any attempt to approach the inner realm of private sexual experience in humans is confronted by three major problems. These may be briefly noted.

A first problem is that human sexual experience explores an intensely private realm, which, one cannot but immediately feel, might well remain so. To inquire into sexuality at the human level seems in some ways almost to intrude upon human nature itself. This reluctance to speak of innermost private experience reflects a strong bent of the human mind. It seems to be part of human nature to keep sexual memories, good and bad alike, in well-contained silence.

This problem must be noted to be one of an inherent subjective reservation, counseling due respect. It is as if the inner mind were reflecting a certain external protectedness which human genital regions externally also demand. If, for valid reasons, an inquiry into the nature of human sexual privacy is to be undertaken, a first qualification would appear to be that appropriate respect for certain inherent sensibilities be maintained. This presents as a major qualification.

A second problem is that much of sexual experience, if not most, characteristically takes place at nonverbal levels. As lovers know, the entire sexual experience can take place silently, and even the fullest experience may take place with but a few words. For much of this realm verbal symbols simply do not exist. In the preliminaries of sexual attraction, in romantic preludes and in exchanges of emotional appreciation, words of endearment are helpful, but once arousal reaches certain levels, words begin to fail. Simple words may give way to more elemental moans or to yells. As this study suggests, perhaps this reflects in part the long evolutionary history of sexuality, which emerged well before words. This is a problem of inherent limitation by the nature of the subject, one much better known to astronomers. Though rendering the exploratory task more difficult, this does not appear to be prohibitive.

A third problem involves unresolved inner conflicts. These arise from inhibitions which have not been overcome and from pains incurred from previous unsatisfactory sexual experience. Each such impediment tends to render objectivity more difficult.

One of the much underappreciated facts of the human experience is that we are all sexual learners. This learning involves profound pleasures as well as experiences which are preferred to be forgotten. When cultural inhibitions are recognized and are properly appreciated to be a hindrance to health or to the individual's development, these can usually be overcome. Emotional trauma from unfortunate previous sexual experience, however, can present a very difficult problem. Sexual actions involve powerful capacities to produce interpersonal injury, which tend to last. Such injuries are often painful beyond a person's immediate capacities for their resolution. As a result, the mind commonly walls them off. Away from awareness, they may seem to be producing no further injury. However, they are far from silent. Unresolved problems from the sexual past tend to render further sexual experiences conflicted and less enjoyable. Repressed processes subtly modify thinking and feeling, even in realms not relating to reproduction itself. In more severe instances, a complete sexual burnout can occur.

The wise counsel, self-honesty and contrite heart needed to over-come well-embedded sexual problems do not come easily, nor does a forgiveness of the self come easily. For these, the only fully satis-factory answer is conscious resolution, preferably earlier than later. Probably everyone has a certain stratum of such memories in various stages of being resolved. To the extent to which sexual issues remain unresolved, however, any study involving human nature tends to suf-fer certain measures of impaired objectivity.

A general cognizance of the remarkable capacity of sexual forces within human nature to inflict injury tends to endow their social per-sona with a certain "across the track" status. These are powers which we know well are there, but which we seldom talk about and prefer often in everyday life not to recognize. One result is that their in-comparable powers for advancing personal growth and enhancing stability at any age tend to be culturally blunted. Although they re-main largely taken for granted, their more integrated and healthfully forwarding cultural use, based upon a sound and comprehensive un-derstanding, remains yet to be fully explored.

A certain note of melancholy may well attend any study touching upon human sexual experience. Perhaps in no other realm of hu-man learning does the reality of experience so often fall below ex-pectation. Yet it is also true that perhaps in no other realm of human experience does expectation so lend itself to renewed engagement with reality. Although problems commonly intervene, humans sense that, fundamentally, their nature is designed for sexual fulfillment. This really is the normal state of a human being, to which all aspire and toward which, this study finds, historically human physiology has directed. Human physiology would not be what it is today had not sexual fulfillment been a recurrently common human denominator.

II

SEXUAL PRIVACY AS A UNIQUE HUMAN NEED

An intense need for sexual privacy is one of the most conspicuous features of human reproduction, and certainly one of the most distinctive in animate nature. Such needs do not appear to exist anywhere else in the vertebrate realm. The intensity of these needs also suggests that they arose by no evolutional accident. Rather, a significant evolutional role is to be anticipated

The human need for sexual privacy is noted to be both universal and strong. Although perhaps for some women the experience of childbirth may compare, no other experience of life appears to be as private as sexual intercourse. Through codes of decency, human laws protect sexual privacy. In almost all cultures incest taboos protect sexual privacy within the family. No subject proves to be more readily capable of eliciting humor than one suggesting intrusion upon sexual privacy. The human sense of privacy is both delicate and demanding. Until the pair become sufficiently accustomed to each other, it takes emotional effort for a man and woman participating together sexually to discuss aspects of the experience. Silence reigns. An act of intercourse in progress promptly ceases when, interrupting privacy, any external alert occurs. Should intercourse then resume, it is seldom quite the same. Wider external realty has shattered a world of inner reality whose complex, elevated state takes time and a certain secure mental preparedness which is not readily restored. Sexual experience has wide meaning for the person, and yet is the subject least likely to be talked about. What, one wonders, could have produced the emergence of such a uniquely human reproductive need?

Until the pioneering investigations of psychiatry were begun late in the nineteenth- century, an objective understanding of the working dynamics of human sexuality remained a realm totally obscure. There was folklore, anecdote and often grievous pathology, but a significant factually based understanding did not exist. Particularly over the past sixty years, clinical and laboratory studies in medicine and biology have done much to put into place an objective understanding of this universally experienced but still poorly understood realm.

Upon scanning the vertebrate evolutional scene, it becomes immediately evident that the human needs for privacy in sexual experience historically are quite recent. Our great ape cousins, the bonobos and the chimpanzees, experience no sexual inhibitions. For these, our closest genetic kin, sexual acts are performed quite openly. Until humans can appreciate that these actions, which exhibit sexuality before needs for privacy had evolved, are essentially appropriate to the genus, their behavior can prove to be embarrassing. Copulation in apes is but mildly personal, at best, and completed within a few brief minutes or, perhaps, even in seconds. The great ape sexual encounter involves a short series of male thrusts. Emotional involvement seems modest at best, of variable excitement, and of short duration. While copulating, an older male may continue to eat his banana. After climax, the participants nonchalantly walk away, each often in a different direction. From a last common ancestor who followed such an open, impersonal sexual relationship, how and why could strong human needs for privacy have evolved?

In facing this question, the relevant distance of the evolutional interval from a last common ancestor of some six to seven million years ago to the modern human can be appreciably narrowed. As to be described subsequently, the fact that sexual intercourse at the human level is uniquely private supports that its specifically human aspects evolved during the last two million years - that is, that it elaborated concurrently with the enormous recent development of the human cerebral cortex. This again means that the basic evolutional context within which needs for sexual privacy arose was that of hunter-gatherer society. This relatively stable period of human protoculture prevailed for the last one and a half to two million years. It persisted for an impressive 99 % of the human career, up to the recent advent of agriculture and the emergence of civilization. This may perhaps explain, in part, the essentially nonverbal nature of sexual intercourse at the human level. The rich vocabulary of language and words of delicate sensibility may well have phased their emergence later rather than earlier in the human career.

A note of caution needs to be made in relation to this study. Its major variables are general; in particular, they are those of general physiology. The advantage of the general statement is ever, precisely, that it is general, capable of serving as a broad interpretive background. Its disadvantage lies in differences with particular circumscribed studies that are likely to have arisen independently or may arise in its use. Where these occur, problems of reconciliation emerge, some of which in the course of resolution are likely to require qualification or restatement of the general.

PART II: THE ANCESTRAL HUMAN CONTEXT

III

THE HUMAN ANCESTRAL TIMELINE

From the time of the departure from a common ancestor with the great apes to today, the human timeline has involved an evolutional interval of six to seven million years. This commenced with the loss of rainforest habitats, coursed through stages of woodland and savannah, and has advanced in moderns to a total habitation of the planet.

*I*n order to understand the role that the advent of sexual privacy played in evolution as a precondition to distinctly human reproductive behavior, it is necessary to appreciate the ecologic context within which it arose.

A modern human today is a product of more than ten million years of evolution that took place mainly on the northern African continent. In the Miocene Era the entire continent was covered by tropical rainforest in which great ape species prospered. In a long, slow, steady decline, however, reduced rainfalls acted to produce such progressive drying that residual rainforest areas today exist only as a wide central equatorial belt. The remainder of the continent has reverted to scattered woodlands, grasslands, bush, savannah, and desert.

The fates of various ape lineages facing this steadfast drying trend have varied. Some simian ancestors descended to full-time terrestrial life, moving about together as baboon troops today. Some species of apes ancestral to modern chimpanzees descended to an edge-of-forest existence, living mainly on the ground but retaining full adeptness for arboreal life when desired or necessary. Some species of smaller primates were able to continue in the trees, thriving on lesser needs for food as monkeys. Surprisingly, one great ape species, the bonobo, our closest genetic relative, was able to continue to live a canopy life in an environmentally limited region south of the river Congo. It is estimated by DNA studies that humans and bonobos separated from a common ancestor some six to seven million years ago.

Recent carbon isotope studies examining the chemical composition of ancestral hominin teeth reveal that, from the level of a last common ancestor with the apes to that of a modern human, the diets of our lineage have coursed through four different ecologic stages. These have been, respectively, the tropical rainforest canopy, woodland, wooded grassland, and the Ice Ages. Although minor variations occurred in each period, the major ecologic features of each are clear.

From the Miocene Era up to 7 million years ago humans shared the same genetic lineage as the modern great apes. This reflected the primordial environment of the rainforest canopy, unmatched in lushness and variety. Temperatures were consistently warm and the atmosphere constantly humid. Foods in the canopy proved to be abundant and varied all year round. Although chimpanzees today may occasionally capture a monkey and eat it, historically the diet of canopy primates consisted primarily of fruits, nuts, and leaves. Here, living at the top of the food chain, ancestors of long ago lived in a world of complete safety; predators did not exist. This was not true of the rainforest's mid-levels or of the ground, regions that were avoided. Distances from one feeding location to another tended to be short, often requiring little movement more than moving a few yards. For the young, freedom of movement was ample, with much room for play. Food loss due to disappearance of the canopy itself, secondary to progressive environmental desiccation, drove our simian-like ancestors from their primal rainforest Eden.

By 6.5 million years ago ancestral species were beginning to search for food in woodland habitats. For simian-like creatures long adapted to canopy life, such environments were new. Temperatures were more variable. The atmosphere was drier and circulated more freely. Fruiting trees, though still available, were less diverse, more widely scattered, and more seasonal in their availability. Instead of ambling from one feeding location to another, the band might now send out scouts. When a ripe tree was found, the rest of the band was called, and shortly all joined in feeding together happily. Often a senior female or male would remember the location of specific trees fruiting at specific times of the year. With often wide food dispersion, appreciable travel would often be necessary for everyday life-maintenance. Similar feeding vicissitudes are noted in chimpanzee experience today.

Although some foods were available on the ground, feeding at such locations remained of low priority. When moving on the ground, the simian-like ancestor, still an arboreal creature, had to resort to clumsy knuckle walking. On the ground such an ancestor was positioned farthest from the treetops where safety could be found. Partly out of convenience and partly in protection from danger, infants and children during ground travel had to be closely attended. Overall, due to limited mobility and to compromised escape, danger was the greatest on the ground.

An ever eminent hazard of woodland life, however, only slightly less dangerous, proved to be the leopard, who also preferred to inhabit trees. Most of the time, this large feline lives on the strong lower branches of larger trees, where he is able to move about with remarkable agility.

When traveling alone on large branches from one feeding location to another, a nursing mother often shares this habitat preference with a predator. If confronted by a leopard, an ancestor moving on a similar branch level could usually dash to higher tree regions where it would quickly lie beyond the leopard's reach.

The main ecologic problem for the band would have been for children, most of all for nursing infants being carried by their mothers. Leopards develop tastes for particular foods, and acquire habits of capture that soon become skills One firm swipe of a leaping leopard's paw upon a nursing infant's body would usually have been sufficient to dislodge it, providing momentary opportunity for the leopard to clamp its jaws upon the child's head and then quickly to dash away. Against a leopard so habituated, a startled woman proves to be no match. With heavy biological investment in each child, the cost of such losses to band and species perpetuation is heavy. A fortunate aspect of leopard life for humans is that, sleeping on large branches during the day, they tend to hunt by night. A second advantage for human ancestors would have been that leopards live solitary lives, rendering confrontation by multiple opponents to prove more effective. Ancestors at this stage would have moved about much, if not most of the time, in groups. Nevertheless, to this day leopards remain formidable predators. We will never know how costly for our ancestors their predation proved to be, but available evidence supports that it is likely to have been quite significant.

By 4 million years ago, under slowly progressive environmental desiccation the woodlands were steadily giving way to bush, savannah, and sparsely treed grassland. As wooded areas became reduced, intervening regions of bush and savannah enlarged. This transition modified ancestral diets, changing them from reliance upon fruits and leaves to grasses and sedges.

Although travel on ground terrain was least desired, there came to be no longer any choice. By this time, however, capacities for walking had dramatically improved, and strategies for travel were able to include the continuing use of trees to provide refuge. When needed for safety or for sleeping at night, all four extremities were used for climbing, with digits retaining their nails and their ancient curvatures, assisting in arboreal adeptness. Much as baboons move about today when on the ground, bands traveled in groups with infants and children protected centrally.

The varieties of predators on the ground were many – lions, hyenas, dogs, saber toothed cats, rhinoceroses, crocodiles, hippopotamuses, and others – but their locations were usually known, and the wide visibility usually afforded by the plains was advantageous. Nursing mothers on open savannah were likely to be accompanied by protective males. Mothers and fathers alike carried their children in transport. It is remarkable how amenable to distance travel a modern human child proves to be. Not unlike the young of antelope species on the plains, a healthy five year old can often walk miles on foot without fatiguing.

The functional ecology of ancestors in the wooded grassland period appears to have remained relatively constant. It is difficult to estimate with any exactness the weight of any single variable in this milieu, but the overall state appears to have remained a context of environmental stability and general ecologic balance. Species that previously had been coursing evolutionally in single lineages were now branching out, diversifying into genetically distinct variations of species.

By 2.5 million years ago this ecologic constancy began to give way to a fourth period. This coincides with the Pleistocene Era, in which we are still living. Cyclically recurrent instabilities were appearing which, in more frequent and severe form, would come to produce the Ice Ages. The first clear landmark of hominin activity in this period was discovered by Louis Leakey, who found clusters of sharp edged stone flakes, sufficient to cut hides and to slice meat. These had been first made by certain late australopithecines. Subsequently, these simple sharp flakes, their cores and related simple stone implements came to serve as the basis for an emergent stone tool industry that characterized the habiline and erectine tenures.

By 2 million years ago, with the Ice Ages well underway, the habilines had advanced to a stone tool ecology, described as Olduwan for the Rift Valley location where they were first found. Once again, early humans were forced into scavenging and into minor hunting for food by desperation during periods of nutritional hardship, now occurring in cold dry periods. Diets now were modified to include meats in varying proportions. From being fructivore and and vegivore, humans now variously advanced to became carnivores, at certain times optional and at other times obligate.

By 2 ½ million years ago hominids had attained remarkable ambulatory capacities, along with the constitutional capacities needed to sustain extended activity on the plains. Emergent scavenging, however, and the wide environmental variations that were coming to be encountered presented novel challenges. New adaptive abilities of diverse kinds were needed. Humans were evolving toward an improved general environmental adeptness rather than in specific adaptations. An australopithecine brain measuring 350cc was little larger than that of a chimpanzee, roughly equivalent to that of its ancestors of 5million years previous. In response to new challenges, brains now began to enlarge, rapidly reaching 700cc capacity in the habilines.

Hunting and scavenging also involved a new way of life, centered about home bases. Whereas previously males and females accompanied by their children had traveled together in search of food, now males went abroad separately to scavenge, leaving women and children behind at safe locations.

As a result of rapid evolutionary development, humans quickly climbed again to the top of the food chain. Although a total freedom from predation was not yet attained – even today a bear can force a human to climb a tree for safety – dangers became dramatically reduced.

Meanwhile, periodic environmental variation reflecting new solar cycles, varying at increasing rates, was becoming more extreme. If organisms are reasonably endowed, a program of recurrent evolutional change can prove to be a powerful engine advancing organismic evolution. If insufficiently endowed, however, severe change can lead to extinction. How could human ancestors have met such long sustained, dramatic environmental change?

Certain physiological features that favored rapid evolution may be noted. Before significant environmental change takes place, organisms must have developed significant metabolic reserves to tolerate the levels of change that are about to descend. When the challenge arises, though incurring new metabolic costs and new inefficiencies, these capacities allow useful adaptations to develop. Given sufficient time, however, particularly an intervening period of ecologic stability, inner reserves tend to restore themselves again to levels of more favorable balance and efficiency. Such inner consolidation is all the more likely to take place when, as pertained during Ice Age history, periods of deprivation are followed by stable periods of ample abundance. Optimal inner stabilities, when restored, tend to remain constant until a new challenge arises, for facing which once again they prove to be favorably prepared. Such fortuitous progression requires that no descent forced by severity takes place beyond the reserve capacities of the system to manage. In the Pleistocene, mirabile dictu, such sequences appear to have recurred cyclically over a hundred times.

Whereas a particular change may first be described as an adaptation, an externally accommodating change, subsequent inner consolidation and improvement tend to produce a more general organismic adeptness, an enlargement of the capacities through which the environment is used for the organism's benefit. Some such serial processes, repeated near to countless times, appear to be needed to account for the unprecedented development that took place in the human brain. This organ attained its modern size of 1,375cc by quadrupling the size of the brain of a common ancestor , enlarging it 1,000 cc in a span of two million years.

Many of the features mentioned in the above highly selective ancestral timeline are elaborated in more detail in the following study. However, before turning to these, a central aspect of the unique human evolutionary ascent that deserves recognition may be noted. This relates to human longevity.

A modern bonobo in the wild, our closest evolutional kin, averages a life span of some 35 years, with an optimal longevity of approximately 50 years. Rarely does an individual life span exceed more than one or two years beyond this. Such viability would appear to approximate that which pertained for of our common ancestor seven million years ago. In a modern human, a comparable life span under favorable conditions reaches approximately 70 years, the proverbial three score years and ten. An optimal human life span reaches 100 years, which can occasionally extend to as much as 115 years. There exists ample room for debate in such numbers where statistical precision still proves to be elusive, but their overall import leaves little room for questioning. In its seven million year history the overall vitality of human beings has significantly advanced. In particular, as measured by comparable lifetimes, human vitality has approximately doubled.

Increased vitality, a dimension of life unable yet to be quantified, is of particular relevance to a major tenet of this study. Reproduction is a major life function that, until birth, reveals little externally of its characteristically heavy internal work. If a female great ape can bear up to 5 or 6 offspring in her lifetime, a comparable human reproductive capacity in a woman more than doubles that of an ape. Such a marked human advance in vitality is compatible with the general proposition offered in this study that the role of reproduction has advanced in humans so markedly beyond simian levels as to realize a new order of functional magnitude.

PART III: BEFORE
SEXUAL PRIVACY

THE HOMINID TENURE

The Ecologic Context of the Hominid Period

*T*he environment of the rainforest canopy, still enjoyed by our closest genetic kin, the bonobo, is benevolently unique. Here the climate is always warm, moderate, and moist. Foods are varied and ample at all times, requiring little effort to obtain in moving from one feeding location to the next. For all great ape species at the top of the food chain in the canopy, predation is absent. This high degree of safety favors an arboreal matriarchy. Group cohesion is often maintained by older females, producing a calm social atmosphere. Should a male in bonobo society become excessively assertive, females promptly gang up upon him, putting him in his place. However, as needs for protection take on a high priority when bonobos descend to the ground, females automatically revert to male leadership and social dominance.

The woodland environment of four to five million years ago, the first stage of transition, has now become well documented. This more open and exposed climate was appreciably more variable, proving both colder during nights and hotter at midday than that of the canopy. The open woodland milieu also presented a change to a less humid atmosphere. As rainfall levels gradually lessened, foods became less abundant, were more variable in their seasonal availability, and proved to be more widely dispersed in their locations. In the woodland milieu hominids had to perform larger amounts of work foraging

over greater distances finding food. An entirely new feature proved
to be that hominids descending into woodland areas, losing the protection of the canopy, became promptly subject to predation. Some
large cats can reach the large lower limbs of forest trees on which
early hominids would also have traveled. Whether or not matriarchy
still prevailed for hominids early in woodland life, the protective role
of the male, critically needed at this juncture, would have invited a
prompt evolutionary emergence of male leadership.

In attempting to envision this first major hominin transition, a preliminary admonition of Le Gros Clark seems to be appropriate. Any
major organismic transition needs to be approached in terms of numerous likely changes of function. Any dominant change will inevitably involve lesser changes, and concurrent changes may occur as
well. While the use of upright stances may have first served to render food more accessible in canopy life, as ancestors left rainforests
the advance to a fully upright gait would very likely have involved
other uses. One, as Darwin suggested, may have been the carrying of
objects, the most important of which, this study suggests, may have
been children. When positioned on the ground, as Leakey and others
have pointed out, the widespread use of the raised head by higher
vertebrates assists in perceiving potential dangers at distances. A sudden noise near to a waterhole can promptly raise the heads of many
drinking animals.

Also, when positioned on the ground, as Leakey and others have
pointed out, the widespread use of the raised head by higher vertebrates proves to be remarkably helpful in perceiving dangers at distances. A sudden noise near to a waterhole can promptly raise the
heads of many or all drinking animals. In triangulation, a small hy-
potenuse in a wafer-thin, angled form can provide perception of the
origin of a base at a surprising distance, a feature commonly used
by resting vertebrates to discern the origin of sudden environmental
sensory change.

In separating from a common ancestor with the great apes, human ancestors departing from rainforest to woodland life began our specific lineage, most of whose member species remain yet to be discovered. The large pools of post-simian species that evolved concurrently in these millennia are described as hominids, while those in direct human lineage, when identified, are described as hominins. If the famous australopithecine Lucy becomes accepted as one of a human ancestral species, her status will thereby change from being hominid to hominin.

IV

THE SEARCH FOR SAFETY

The loss of rainforest canopy security introduced human ancestors to a long evolutionary transition involving widespread new susceptibilities to predation.

*A*s ancestral hominids became required to venture more and more widely in the enlarging woodland milieu in their search for food, the major challenge that arose, as noted, was not one of sufficient food or the ability to find it but new dangers of predation. Whereas in forest canopies great apes live in almost total freedom from predators, in woodland environments this dramatically changes. No great ape is a match for a large canine. A mother carrying a nursing infant proves to be a highly vulnerable target. Except in desperation, a leopard normally will not seek a fight. However, a hominid female less than four feet tall, with bent posture and shuffling gait carrying in her arms a nursing infant, presents little challenge. Infants are always choice morsels. With a brain crafted for stealth, the leopard suddenly lunges forward upon a startled mother, causing her to lose her balance and momentarily to lose her grip. A quick clamping of the leopard's jaw upon the infant's head or neck is followed by a lightning dash away, clutching its prey.

Paleontologists are appreciating that such threats to ancestral hominids forced out of rainforest canopies persisted for many hundreds of thousands of years, indeed, throughout most of human prehistory. The very first skull ever identified as human, that of the Taung child, showed large claw marks indenting its cranium. Many hominid fossil bones, including those of adults, show canine tooth marks, indicating that these individuals had served as prey. Even the digits of the first human, *Homo habilis*, show curvature, indicating the necessary recourse to the climbing of trees for safety.

The confrontation of early hominids with predation was but the beginning of a long evolutional contest that is not even entirely over today. The leopard appears to have been the main predator. It is sometimes held by ecologists that leopards and humans have evolved together. Any farmer walking home at night on the African plains still has to be prepared to be confronted by a stalking leopard. Many an African community inexplicably loses members until it is discovered that a local leopard, hunting by night, has chosen humans as its preferred kill. Once a particular leopard's feeding tastes have been acquired, they tend to persist as habits. Before their cause has been discovered and corrected, such losses have been known to reach human body counts in the scores.

Early hominids appear to have been forced into close challenge with the leopard by their habitat change. Although the leopard mainly hunts on the ground nocturnally, its preferred habitat is in the trees where it likes strong lower limbs on which to rest. There it drags its kill to be eaten, there it is comfortable, there it moves about with dexterity, and there by day it sleeps in safety. Into this lower arboreal realm human ancestors, preferring to avoid ground travel, were forced to move by the progressive loss of the canopy habitats to which their bodies were so well adapted.

If it is maintained that humans and leopards moving about on lower woodland levels evolved together, the most intense period of such interaction would appear to have been the first two to three million years after departing from a common ancestor with the apes. By the level of ardipithecus, four million years ago, ground adeptness had so advanced that both arboreal and ground travel were well in place. A clear lumbar lordosis in adult males and females alike indicates the habitual carriage of children by adults during travel. This suggests the attainment of moderate levels of ground safety. At this later stage larger varieties of predators had to be faced. However, by this point conflicts appear to have become less costly to the species.

How might such a new challenge have been met? The first defense in woodland and ground locations would have been to move together in close groups. This was the course taken by ancestors of the baboon upon their descent to the ground and by the common chimpanzee. Even the famous Lucy, an australopithecine living three million years later, moved in a group of approximately 30 individuals. In such bands younger adult males providing general defense move about at the periphery while older males move centrally, guarding females and children.

Even in group formation, however, the predatory challenge remained significant. The leopard carefully observes its potential prey, searching for the weak and the vulnerable. It bides its time until the chosen kill is separate or in some way disadvantaged, and then acts with matchless speed. The mere impact of such a hurling mass would prove to be quite sufficient to topple any hominid, whether it be a man or a woman, alone or carrying an infant. This says nothing of the sudden fear or pain from claws digging into the mother's flesh. For the leopard, however, when hunting on the ground such a well orchestrated catch is but guarded sport. Certain hunting habits of the leopard would have favored human ancestors. The leopard is a night hunter, by which time most ancestors would have become safely located in their tree top nests. Traveling at dusk, however, would have been particularly hazardous.

From the standpoint of a victim, however, each hominin lost, a large biological investment, proves to be costly to the reproductive status of the band. Where each offspring represents such a high biological investment, too many infant losses prove to be incompatible with a species' survival. Perhaps human ancestral species were able to survive only because new dangers of predation came on slowly and occurred only sporadically.

The unfolding paleontological record indicates fairly clearly that the upright gait of humans commenced during this time. For the first 3 to 4 million years after leaving forest canopies, improved ambulatory capacities, new abilities to walk and to run, dominated the evolution of our ancestors. Body balance would have become more important. Brain capacities were apparently quite adequate to the general level of ecologic need. Postural erectness would have helped to detect feline and other dangers earlier. Perhaps improved safety was one of several essential functions which benefited from erectness. Upright gait would also have helped in acquiring less accessible fruits or nuts from trees in the dry season, as well as in carrying various other objects, particularly infants during a travel period.

A study by Naessen and others on modern woman's sense of postural balance offers suggestions which appear to be particularly relevant to this period of early hominin descent. Falls due to deteriorating postural balance in postmenopausal women commonly result in debilitating hip and spinal fractures. Devising a tilt platform by which to measure body balance through measuring sway, Naessen's group found that a woman's capacities to maintain balance have become so depleted by menopause that relatively small amounts of estrogen therapy given at this time prove to be able to restore capacities for balance to levels normally seen in young women. This impressive study, subsequently well confirmed, suggests that a surge of estrogenic activity early in the hominid career served to aid the clumsy ambulatory ways of these ancestors. This particularly pertained for improving the balance of mothers trying to knuckle walk on large tree limbs while carrying nursing infants. Of all possible situations endangering an infant, this would appear to have been high on the list.

An estrogenic surge at this early hominin level would have had several other related consequences, many to be described subsequently in more detail. One would have been that woman's weight-bearing bony structures became strengthened, aiding her capacity to carry infants and small children over distances. At some early date, estrogenic advances in conjunction with those of other hormones would have contributed to maintaining permanent breasts in adult females.

In great apes breast formation takes place only postpartum while nursing, regressing afterward. The emergence of permanent breast development in humans permitted infants to suck at any and all times desired, and thus to remain contentedly silent while in transit, even long after needs for nursing itself have been met. Whereas the sucking reflexes of great apes are used almost entirely for nursing, these reflexes in humans have become markedly extended. Many a child of four or five years of age still sucks its thumb. The combined effect of permanent breasts and lengthened sucking reflexes permits infants to continue sucking while asleep and being carried. Among other things, this served to prevent crying out, thereby giving notice of their presence to nearby predators.

What was needed foremost in this new environment, however, would have been for the male to be brought into the picture, particularly to protect any woman with a child. Such a role, now more critical, would not have been unknown to simian species. When chimpanzee youngsters are at play in a clearing on the ground, a senior male commonly watches over them. Similarly, whenever the women, children, or young of a species of large monkey leave the forest to raid fruit trees on the ground at the open periphery, a feature often observed by the writer, an older male remains behind watching the group from the treetops. At the first warning of possible danger, a deep throated call from him sends the entire group promptly running back to the forest for safety. In all great ape groups a protective role of males is almost universal. Exceptions do occur, as when males attempt infanticide, apparently trying to bring on estrus in a lactating female. At such times, locating themselves at the band's periphery, mothers guard their young until they are larger and less vulnerable.

Evolution's method of drawing the male into the picture for improved female protection, apparently, was twofold: first, to advance vaginal function to become potentially accessible at any time when it might be advantageous; and, second, to lengthen the duration of intercourse. The first development constituted an advance from cyclic sexuality to continuous, while the second, apparently at a much later date, would appear to have served to increase the forces of interpersonal affiliation, eventually producing pair bonding. The latter development was central to specifically human evolution. This latter advance would appear to have been central to the evolution of subsequent humans, *Homo erectus*. Such major evolutional advances must have involved many changes in numerous organs over an extended interval of time. At this evolutional stage, increased vaginal access would appear to have been central to the further evolution of hominid sexual elaboration.

As all women know, acute fear or intense anxiety can terminate menstrual cycles at any time. However, hominid fears most of the time could not have been this severe, or the reproduction of the entire group would have failed. The fact that the leopard is a solitary hunter suggests that successful predatory episodes, though they are quite costly, proved to be reasonably sporadic and were frequently able to be fended off or avoided. Only at a later evolutional date with more extensive ground living did a much wider variety of potential predators, ranging from crocodiles and hyenas on the ground to hawks overhead, enter the picture.

While the need to terminate periodic estrus became imperative, other actions of increased estrogen production which may have made their evolutionary entrance at this point may have helped in other ways. It is a common feature of endocrine evolution for hormonal actions to progress from intermittent and periodic roles to participating as steady components in more advanced hormonal programs. If so, instead of being present for approximately five days a month, vaginal access through increased estrogenic action now became potentially available for larger numbers of days, and eventually during the entire month. Evolutionally, this suggests a general physiological advance of estrogenic actions from first serving in a specialized cyclic or seasonal reproductive role to participating in everyday maintenance. Whereas in lower vertebrates estrogen serves to activate reproduction in season, in humans it advances to prepare tissues for reproduction, performing a sexually priming role on a continuous basis.

If estrogen were to have advanced at more developed hominid levels to become sexually a more highly activating agency, such a change would have gone against the emerging ecology of these species. In the new milieu where predation could occur at any time, a heightened or an unrelieved sexual drivenness would have tended to invite disaster. Rather, awaiting appropriate activation by other agencies, estrogen serves to maintain tissues in reproductive readiness, a role which has persisted through all later human evolutionary levels up to and including moderns. The role of estrogens in human physiology appears to be one purely of priming. In the present study, the emergent hominid sexually activating forces would appear to have been emotional and tactile processes, brought into action on appropriate occasion. This permitted copulation to advance from being an action rigidly programmed by the drive of endocrine cycles to one able to be entered into according to voluntary discernment and situational appropriateness.

In a most general perspective, the evolutionally rapid rise of safety needs in female hominids, emerging along with woodland occupation, may have served to elicit certain accessory cognitive needs relating to male protection. For fertile women, larger and more active males, as well as those more inclined to interpersonal attentiveness, may have begun to have an inherent sexual attractiveness.

Varied safety needs are far from unknown among primates. These differ from species to species and from situation to situation. A certain baseline of mammalian safety needs can be observed at the level of the prosimians. In lemur ecology, female safety needs are usually well met by males. Daily, the male patrols his home territory, which includes the smaller living areas of several females. Upon completing his patrol, he returns to his central nesting area, greeting each female in turn in her nest. A warm hug reassures her that he is still present, protecting her and defending her small feeding range. Should no male be present to give this assurance, however, perhaps having been killed, the female relocates her nest. She moves on until she finds an area where a male will attend her. Should the male happen to find her in heat, he remains with her in the nest until her sexual receptivity is over, a brief period which usually lasts but for a fraction of a day.

At this level, the hug, inherited from reptilian reproduction, begins to show elaboration. While it continues to be used in copulation, it has developed to become a broad gesture of protection. It will remain this as it further evolves in humans to serve as a gesture of friendly greeting, joy, triumph, or simple appreciation. Human uses range from the reassurance of a frightened child to the shared exultation of team players enjoying their victory in a football match.

Because the female after giving birth, in order to nurse her young, needs to marginalize herself from the general arena of ongoing group activities, a stratum of safety needs tends to exist as an aspect of her gender nature. In this broad perspective, female safety needs are part of woman's inheritance; their existence subsumes a complementary existence of adequate male protection.

Whereas the above considerations suggest that safety needs became acute early in hominid history, by the level of humans at hunter-gatherer stages such dramatic advances in environmental mastery had been attained that safety needs had markedly receded from everyday awareness. Humans had moved again to the top of the food chain. The much greater security attained by humans in civilized cultures has permitted safety needs to recede yet further, to near obscurity. In most human societies today, most individuals most of the time move about freely, with little or no awareness of needing any male protection.

Almost obscure, and yet not entirely. In human females such needs have not completely ceased to exist, but, rather, have become marginalized. New levels of safety dampen them. However, many women today, perhaps even most, continue to have such needs, well ingrained, existing largely at levels below usual awareness. When they assert themselves in awareness, such needs, retaining their inherent prepotency over sexual needs, tend to detract from a woman's full sexual responsiveness.

In almost all women a silent stratum exists desiring to be held as a preliminary behavioral event before proceeding in intercourse. Such needs do not appear to exist in our closest relatives, the great apes. However, safety needs when they become strong always act with prepotency over most other needs. In any sexually reproducing animal, these needs are always required to be met before the sexual encounter can proceed. Their often potent presentation in modern woman's nature today may reflect the profound hominid sensitizing experience with predation. This began with the departure of primal ancestors from the rainforest and did not finally come to an end until, some four to six million years later, ground mastery had been attained.

As with so many other needs, those in a woman desiring to be held by a man vary enormously from person to person and from time to time. Marriage counselors often note, usually in passing and deferring to more pressing concerns, that a woman who is happily married may comment that her husband makes her feel safe. Many a married woman cannot sleep on the nights when a husband is away. Some women barely desire to be held, if even at all. For certain others, however, such needs can be so strong as to comprise the major satisfaction of a sexual experience. Indeed, it is reported that in some women these needs are so powerful that, when unmet, these persons may even resort to prostitution to have them satisfied.

It is often stated by comparative physiologists that the best record of an organism's past evolutionary history lies its nervous system. Where gender relationships are present, male and female need to be considered in relation to each other. A modern woman's inherent needs to be held may well reflect, at least in part, the ancient necessity of overcoming predatory dangers that repeatedly had to be met for more complex hominid reproductive processes to proceed. Such womanly needs are complemented by male needs. Many men experience a strong innate desire to hold a woman affectionately, vaguely sensing the presence of a subtle dimension of necessary protection, even if the immediate civilized environment, in its removal from so many dangers, typically has no occasion to elicit it.

The need in a woman to be held, and by a man to hold, is complemented by an interesting, almost universal psychological matching at the level of humans, which also relates to basic paired security needs. A male personality trait which women tend to find highly attractive, both consciously and unconsciously, is self assurance. Through a comfortable sense of achievement, demonstrated often by good natured self-confidence, the human male, in essence, states to the female that many bridges toward maintaining a fully adequate general competence in life have been successfully crossed. Upon such a reassuring note of safety a woman senses, often unconsciously, that much, perhaps even most of her own prospective happiness depends.

V

ENTER EMOTION AS A SEXUAL MOTIVATIONAL ADJUNCT

As estrus waned, emotional processes became engaged as an aid in sexual arousal.

A major accessory force in the transition from monthly estrus to states of continuous vaginal accessibility would appear to have been new developments in emotional processes. Described physiolgically, emotions comprise systems of accessory energies which are available for any important need in a vertebrate organism. Such a relatively rapid transition of a major biological function as loss of estrus would have presented an ideal invitation for emotional development. Emotions develop first as undifferentiated excitement. This proceeds to differentiate into pleasant, attractive elaborations - positive states - or to unpleasant, repellant negative states.

In its main features, a likely sequence in hominid emotional elaboration appears able to be envisaged. Upon departing from rainforest safety, as noted, a woman may experience anxiety or fear in her widening search for food, particularly if she is carrying a nursing infant through potentially dangerous territory. Should a male attend her in transit, she cannot but feel grateful. While traveling, she cannot return such a man's favor. Even if she tried, her level of anxiety or fear might well prevent it. However, once located in safety, perhaps enjoying fruits or nuts with others in a relaxed or even jubilant atmosphere, her fears would recede. Her mood would give way to an enhanced appreciation amidst group exultation. Great apes commonly express shared joy when feeding together in abundance and safety.

What is suggested is that a new range of emotional supplementation developed fairly quickly, realizing situational sexual availability. The estrogenic advances that were taking place at his time would have tended to increase the tissue responsiveness of the female genitals. Voluntary stimulation by males at feeding sites after close accompaniment during travel could tend to elicit localized autonomic action. Such voluntary arousal would mark the point where, beyond the fixed rhythms of estrus, women began to respond genitally through emotions that were aroused by a desired male. Such a capacity for sexual response outside of estrus is distinctly human. For this to have commenced early in the hominid career would have been doubly appropriate, marking alike the first appearance of a distinctly human coital feature and introducing new capacities for volition to modify more instinctive behavioral patterns.

It is important to appreciate, however, the nature and limited level of emotional development which would have been possible at such an early stage. On the one hand, varied interpersonal affinities, some permanent and some transient, always exist among the members of a primate band. These would lead mildly toward pair preferences for accompaniment in transit and for copulation at its end. The gamut of emotional responses observed at great ape levels is essentially the same as in humans – perhaps a dozen or more distinct states. Their expression is intense. Lacking in apes are emotional control and fine differentiation. The emotional involvement in such pairs, though often intense, would fall short of permanent interpersonal affinities such as are mediated by higher brain development. As in the bonobo and chimpanzee of today, copulation at this early level would remain brief.

Genital organs would not have evolved to levels sufficient to sustain full or permanent pair constancy, but, rather, would have been commencing so to elaborate under new endocrine development. Rather, a sexually fluid situation would tend to prevail in which males have more frequent but still brief experiences of intromission. Increased rates of copulation in early woodland stages would very much reflect alike the attentive actions of males toward women while moving about from one location to another, as well as the ever changing fluidity of interpersonal relationships within the band.

Along with arousal produced by emotional processes during significant woodland mobility, a rise in testosterone in both sexes may have helped to contribute to the emergence of situational sexual activity. Testosterone is a relatively short-acting hormone which increases with striped muscular activity, as well as with mental attitude and immediate experience. A human adult watching a baseball game is found to experience elevations of testosterone when the team wins and reductions when it losses. The striped musculature of the body which negotiates walking is both physiologically enhanced at any given time by the body's existing level of testosterone and tends to produce additional testosterone during muscular activity. The latter is an example of the body's tendency to facilitate behavioral activities that are in progress. Male testosterone levels after intercourse tend to be higher than before, as may be noted by deepening and relaxation of the voice. Testosterone production in males is characteristically far greater than in females. Thus, at the end of a journey, the male is likely to experience some increase of sexual need, which at this evolutional juncture would have been both opportune and appropriate. If similar effects take place to any extent in the human female, such stimulation would also have assisted her sexual arousal. The one caveat here would be that neither the male nor the female be too tired.

It is quite possible that a phase change in regard to the initiation of sexual activity took place early in ground descent. With decline and loss of estrus, levels of sexual desire in the female would have markedly decreased. Her sexual desire now came to depend upon the actions of other agencies: emotional processes, personal appreciation of the male, and tactile stimulation. On the other hand, with testosterone levels already physiologically much higher than those in the female, males came to experience increased testosterone production from ground travel. During travel males, too, would feel anxiety and fear. However, upon reaching their goal, with newfound safety and perhaps accompanying crude outbursts of female appreciation, it would not take long for circulating testosterone to express itself. In general, from this evolutionary point onward, the stronger sexual initiative would begin to lie in male hands. Among other factors, cognitive assent by the female permits the male to commence tactile arousal. However, the new plasticity of human behavior which was beginning to evolve with terrestrial activity, and with extended walking in particular, would also leave the sexual initiative open for the female to use at any time her needs and judgment might so incline her. Such an inner phase change relating to male initiative in eliciting sexual intercourse at an early woodland stage would be compatible with the presumed new external male role assuming major group protection.

While improved interpersonal affinities were apparently emerging in early hominids, two features would have worked against constancy in pair bonding. One would have been that the dangers of predation would not have permitted the sexual act to be prolonged; rather, it would have to remain what it had been historically - as brief as practicable. Even such high vertebrates as zebras and wildebeest remain never more vulnerable to predation than when mating. The other consideration would be that band ecology had not yet advanced sufficiently to incorporate a total pair constancy. In any higher primate band adult females typically outnumber males, at times by large margins. Simply for the band to survive, some males are required to attend more than one woman with one young child or several. Under severe environmental challenge, however, male capacities for protection can come under such strain that all reproductive concerns become prohibited by the mobilizations of stress.

A third consideration may also have been relevant. Emotional processes were evolving, but the hominid brain, weighing still that of a modern chimpanzee or bonobo, had not yet reached levels of such advanced interpersonal sensibilities as can sustain lasting pair constancy.

Testosterone is a difficult player to envision evolutionally. It probably advanced by multiple stages during the long hominin career which eventually included the human, but would be difficult to pinpoint regarding its level of engagement at any given stage. Furthermore, its evolutional advance in the hominid and human careers would appear to have been greater in the latter, specifically human tenure than in the former. What appears of foremost relevance appears to be that in both tenures all reproductive hormones made certain advances, with estrogens elaborating eminently in the earlier, hominid period, and the testosterone complex advancing predominantly in the human.

What stands out in these considerations appears to be the fact that, once security and safety needs have been met, emotional processes tend to emerge as basic for female sexual arousal. Whereas sexual arousal in the male tends to be supplemented by excitement, the primal emotional state, in the human female a more complex set emerges. This first relates to protection and secondly to a primal liking of the person. When these preconditions are met, the female permits a favored male to make tactile advances. These fundamental priorities appear to persist, recurrently differentiating as higher evolution takes place.

While emergent emotional processes appear likely to have played the major role, and while certain actions of subsequently produced hormones may also have helped, a third major player affecting arousal in these emerging evolutional contexts would appear to have been a new range of tactile processes.

VI

ENTER TOUCH AS A
SEXUAL AROUSAL SYSTEM

With loss of estrus, human ancestors enlisted the
sense of touch as an aid to sexual arousal.

*W*ith the slowly forced transition from rainforest canopy to woodland environments progressing steadily, it became ever more urgent that the group disorganization produced by estrus be terminated. Such behavior became increasingly dangerous. Although termination as soon as evolutionally possible became expedient, few challenges prove more difficult to change than one, such as estrus, deeply rooted in a species' physiology. Although emotional processes would have been immediately mobilized, more supplementation would ultimately be needed to replace the powerfully ingrained phenomenon of estrus. It would appear that touch, widely used for shared grooming, now became employed heavily in focal pairing as an aid to arousal.

Evolving nature appears to have solved the crisis of estrus not by compromise or by any form of regression but by the realization of a major evolutionary advance. In one sense, perhaps, these early ancestors had no choice. Estrus could not be removed; it could only become somehow superseded. The superseding of estrus, however, came to involve the elaboration of a new reproductive framework for the human ancestral lineage.

That such an advance could take place at all speaks well for the physiological status of the last common ancestor and its early evolutionary descendants. Ages of life in the rainforest canopy had apparently produced organisms with high physiological reserves, fitting them well for new adaptations in woodland and in later savannah.

As in any evolutionary transition that is successful, during this initial period a basic level of species reproduction would have had to be maintained. Although the details of the transition from estrus to reproductive organ constancy must remain conjectural, a likely basic sequence seems to be discernable. To whatever level either emotional or tactile arousal came to be incorporated, the intensity of estrus would correspondingly have been able to become reduced.

Greater capacities for tactile arousal in early hominins appear to have emerged in several ways. As noted, all species of primates spend hours daily in mutual grooming. This may take place in twos, threes, or in small groups, all of whom quietly groom one another. This practice produces a state of inner calm and comfort. Although some selection of partners often takes place, in general, loose bonds of affection are formed by widespread grooming. The most powerful bonds develop between mothers and their offspring, which remain strong for life. Grooming is commonly engaged in to reduce hostilities or to form social alliances.

Grooming appears to be the major force holding the great ape band together as a group. In Jane Goodall's studies, a subgroup of her band broke off, feeding within a portion of the original band's territory. Later, warfare broke out, with the larger group completely exterminating the smaller. To Jane and others who had known these chimpanzees for years, once all so playful as a natal group, such an eventuality previously had seemed inconceivable. All were left stunned. How could this have happened? In reply, several things were noted. First, the breakaway group was laying claim to a portion of the band's territory, with its foods. In bad seasons these could prove to be essential for surviving. Second, the members of the seceding group were no longer sharing affections through mutual grooming. Members of each separated group became perceived simply as moving objects, each group to the other. Grooming appears to have once served to hold the members of the original band together.

What appears to have taken place among early hominids with apparent evolutionary rapidity was a behavioral transfer from customary group grooming to more selectively paired grooming by gender. As the skin became more sensitive to touch due to increased estrogenic action and genetic advances, male contact with the female's skin, reinforcing affectional processes, proved to be able to contribute to her sexual arousal. This skin contact was not the firm press or hold of a hug but the gentling of light touch, that of grooming. Humans appear to have evolved an entire neural subsystem of light touch which does not appear to have been equally developed in the great apes. With sufficient touch on skin surfaces, the circulatory perfusion of the human skin increases, increasing its sensitivity. Increased perfusion on the face and chest may be noted as a sex flush, which can also be elicited by emotional processes.

Sustained summations of light touch in conjunction with warm emotional feeling would appear to have combined to elicit a new vaginal access. With continued stimulation and warmly engaging interpersonal behavior, vaginal tissues could swell and increased secretions become produced. In combination with the possibly of faint androgenic action upon her vulvar tissues, the female became able to welcome the male sexually at any desired time. Likely occasions would probably have arisen fairly often, as upon a journey's accomplishment, after traveling together, having found food and chatting amiably together with exuberance in a safe location. Before our more recent ancestors discovered the cooking of food, earlier ancestors had to spend much of the day finding food to eat. At the woodland stage, travel for food became a way of life. This presented a major change from canopy life.

For higher primates the state of grooming is far from the trivial event which it may at first seem to be. Rather, it is one of basic satisfaction, of social pleasuring and visceral contentment. In many ways it presents a base reference for higher states of inner feeling. This state in satisfied simians is the evolutional antecedent to what later comes to be at hunter-gatherer levels a sense of home. It would appear to be no accident that sexual activation tends to emerge in such pleasant and comfortable settings. Emerging at the level of the hominids, paired grooming, producing states of prolonged contentment, came to involve paired relationships, upon the basis of which needs for couple privacy could subsequently emerge at more developed evolutional levels.

Certain features of the human sense of touch may have made their evolutional debut at this level or become accentuated. The sense of touch discriminates between persons one likes and those disliked, a perceptive quality unlikely to have advanced to any extent in ape settings, where universal comity is needed. If a person is liked, touching of any sort is welcome. If a person is detested, an inner voice says, "Don't touch me!" In advancing to make such a distinction in feeling, evolving frontal lobe processes were anticipating the responsibilities of parenthood.

An interesting difference may be noted between the touching of oneself and being touched by another. The central nervous system gives full discriminatory regard to touch by oneself anywhere on the skin. However, touch produced by another person appears to engage the inner reward system to far greater extents, rendering it to be preferred. Thus appears to have developed at hominid levels a usually significnt desire to receive touch from others, producing, in time, an affinity for experiences of personal closeness.

As frequently occurs in evolution, important changes which are interrelated often take place concurrently. New sensibilities in the more vascular female skin appear to have been accompanied by loss of body hair, or, more precisely, by a selective abiotrophy of the large body hairs of apes. Excepting those on the top of the head which serve to protect it from direct sunlight, and those of the genital and axillary areas which prevent chafing, most of the large hairs of the body in hominids proceeded to regress to such diminutive size as to seem on gross inspection quite to have disappeared. Their numbers in humans today actually remain approximately the same as in great apes. Although almost invisible to the naked eye, by their greater incorporation of skin contact in their activation these abiotrophic hairs prove to be more tactily sensitive.

Evidence supporting that such hair "loss" took place at the woodland stage of hominins comes from three sources.

In an elegantly detailed, now classical study of human thermodynamics made over 40 years ago, Newman maintained that, for reasons unknown, human hair loss must have occurred at a woodland stage. In the forest canopy total body hair in human ancestors, as in all primates, was necessary to protect against direct sunlight. Furthermore, since the ambient air was so humid, the sweating of bare skin would have been unable to be used for evaporative help. On the plains, on the other hand, where every other animal of any size is also protected from direct sunlight by ample hair, hair on human ancestors, had it been present, similarly would have been protective and would not have regressed. This did pertain for hair on the scalp. Human hairlessness on the savannah is far more a physiological handicap than a help, even though increased capacities for sweating are present which would help in a chase. Thermodynamic considerations support that human hair loss must have taken place in an environment reasonably protected from direct sunlight and in which open circulating air could make full use of evaporation for the body's delicate temperature control. Though the reason for this change was unclear, Newman postulated that this would have had to have taken place in an openly wooded, parkland environment.

A second line of evidence comes from more recent genetic studies of human lice. Body lice were transmitted to human ancestors at a date estimated to have been over 3 million years ago. Once on the human body, some migrated to the head, becoming a species of head lice, and others to the genitals, becoming a species of pubic lice. Studies of these species are interpreted to support that human hair loss took place before 3 million years ago. This dating also precludes the possibility that human hair loss took place primarily for sweating as an aid to running and hunting on the plains. This appears to have been a secondary benefit of later evolutional use.

A third line of evidence is suggested by a trip to the beach on a summer afternoon. There, white males are noted to be the only humans with traces of long hair anywhere on their bodies except on their heads. It is a comparative anthropologic fact that pure black males in central Africa are hairless except for that on their scalps. The same is true of the pure Chinese, and was true of the pure American Indian. An occasional white person can also be found without hair on the trunk, though not common. The white males showing long body hairs appear to retain residual remnants from an ancestral long-haired state.

Of particular interest, however, are the patterns of residual hair. Hair can be quite profuse on a man's upper chest. This growth extends from one breast area to the other, either isolated or fully developed between the two. This crop tends to extend distally, tapering to the umbilicus or below. In some men, a prominent triangular patch, often alone, exists on the lumbosacral shelf of the low back. Various other hair patterns may be present, such as suggestions of epaulets on the shoulders. However, there are also some areas where long hairs are universally absent. These are located particularly on the sides of the torso and the pelvis.

Most physicians are aware that many women have residual long hairs in the periareloar areas of their breasts. Often, for cosmetic reasons, they wish to have these removed. Also, an occasional patch of small-ish hairs is often noted on the lumbosacral shelf of a white wom-an. This is similar in area and design to that in a male, but far less developed.

What these features of residual long hairs suggest is that at an ear-lier period in human evolution these hairs were so important that they became unusually well endowed genetically, becoming longer and stronger. The residual hairs about a woman's nipple areas suggest that, as mothers carried their infants in walking position, the child held onto the chest by grasping a clutch of long hairs with its strong grasp hand reflex. The residual long hairs on the man's trunk suggest that men also carried children while traveling, particularly, it would seem, those over a year old. An infant could hold onto the long hairs of a man's chest between nipple areas, just as it had on its mother while nursing. Apes grasp with both hands and feet. A slightly older child on a man's back or a woman's could grasp the little triangular patch on the lunbosacral shelf firmly with its toes, while putting its arms on the shoulders, around the neck or holding the man's long head hairs. By holding onto such long hairs, a human child, much like the child of a great ape riding its mother, could ride on its father's back during travel. However, where it would have been dangerous for a child to hold, lest it fall off, would have been on the man's sides. Evolution appears to have solved this problem, attained no doubt at a cost, by avoiding long hair growth on the sides of the torso and pelvis.

Recent findings of the fossils of *Ardipithecus ramidus*, a possible human ancestor of 4.4 million years ago, support that, as the ranges for finding food and maintaining safety widened, the carrying of infants and the young by mothers and fathers alike became common practice. A careful analysis by Lovejoy of the spinal column of "Ardi" reveals a lumbar concavity in males and females alike that is absent in great apes and which, he holds, would not have been present in a common ancestor. This he interprets to be an independent development that occurred early in the hominid lineage.

The present study suggests that the function of the hominid lumbar concavity, serving to form a shelf, would have been to serve as an aid in the carrying of the young. By grasping onto the parent's long hairs in this area with their strong plantar reflexes, their highly compromised position during travel would have become more secured.

These considerations suggest that the primary reason for human hair loss was an evolving urgency for the carrying of infants and young while traveling, particularly when moving about on the ground, and that this took place at the woodland and parkland stage. Apparently australopithecine mothers, continuing the practice of other simian species, carried their infants and young during transit. Moreover, extending this practice out of sheer necessities for safety, adult males appear also to have become involved in this functional advance, perhaps carrying those children a little larger than those carried by mothers. The most appropriate time for such a general hair regression to have advanced to produce a complete functional abiotrophy on arms and legs, and a near to total abiotrophy on trunk areas, would appear to have been at the early woodland stage, as hominids first came to be confronted by various predators. Apparently, both mothers and fathers carried infants and young during transit for many years. The burden of carrying, however, does not appear to have been divided equally. When traveling from one location to another, it was usually essential that males take the lead. Theirs was the foremost responsibility for protection, and theirs usually was the greater familiarity with the territory. In keeping with a general division of labor between the sexes during ambulation, women were the preferred carriers of objects most of the time. Although this division commenced early with carrying nursing infants, as hominin evolution progressed it came to include gathered foods carried in skins, pitchers of water carried on their heads, firewood in bundles, and other items. That women carried the larger burden of children is suggested by the fact that lumbar lordosis is more marked in women than in men today. It would usually be likely that a youngest child, still nursing, would have been carried by its mother. If a next to youngest were needed to be carried, this second burden would more likely have enlisted the father's help.

This reading of the record appears much to oversimplify what must have been a long and frequently changing evolutionary course. In particular, the residual hairs present in white males and females today would appear to represent a second growth of long hairs that previously had completely regressed. In a broad interpretation which includes other facts, as evidenced by blacks in Africa, the pure Chinese and an occasional white male, all humans of all types appear at some time to have "lost" truncal hair. Most white females remain essentially abiotropic, hairless. However, some white human migrants who became trapped in Ice Age environments would appear to have commenced to regrow the ancient hairs previously regressed. Besides the regrowth above noted, white males are noted to have a fine hairy regrowth on the arms and legs in those distal areas most exposed to cold, as well as in that exposed to cold most of all, the face. These areas of lighter regrowth terminate irregularly, tapering in about mid thigh and lower arm regions where the protection of clothing would commence. Most white females, far less exposed to Ice Age cold, show only slight hair recurrence. This occurs mainly on the arms and legs where, in a person wearing body covering, the extremities are most exposed to cold. Such recurrent small hairs on their legs, when present, are considered unsightly and are frequently shaved.

At some time, in sum, the evolving hominid reproductive economy would have benefited by the general baring of the body surface, providing access to a new level of tactile stimulation as a further aid to sexual arousal. This may well have happened early in the wood-land stage, where it would accord with other advantageous changes taking place. As hominids stopped to feed at new locations, males and females, feeling sexually desirous, may have turned first to the relaxing introductory pleasures of grooming. Such a development by its nature comprises a major step of volitional advance in human re-productive processes. At the same time, the shift from group groom-ing to an increased focus upon couples presents a first step toward the creation of sexual arousal through paired stimulation.

Of all of the human senses, touch is potentially the most powerful. Emotional processes are at once omnipresent and also, potentially at least, of incomparable potency. In engaging emotion and touch as ad-juncts in sexual arousal, early in hominid experience human evolu-tion appears to have recruited the most powerful accessory resources then available in evolving human physiology.

VII

THE ECLIPSE OF ESTRUS

> With the change of habitat from rainforest canopy to woodland life, emergent emotional and tactile processes enabled estrogenic advances to replace estrus with apparent rapidity.

*T*he considerations explored in this study suggest that the primary force leading to the eclipse of estrus proved to be a signal environmental change. This involved a transition from rainforest canopy life to woodland habitats. However, many essential forces appear to have been at work. A certain amount of comparative background at this point may help to present a broad perspective for understanding this major reproductive transition.

Although a female bonobo in estrus, our closest genetic kin, may disrupt the social life of the band, in the rainforest canopy this matters little. For approximately two weeks in a 45 day menstrual cycle this female exhibits her swollen, red vulva. No sight is more exciting to bonobo males. At the same time, her waxing and waning vulvo-vaginal swellings drive the female to repeated mating activity. In any presenting bonobo female, this agitated sexual state likely to lead to copulation lasts for approximately a week.

Although in a presenting chimpanzee, our next closest kin, estrus lasts for only approximately five days, chimpanzee bands on the ground experience much greater group disruption during the days when a female is in heat. An excited state prevails for the entire band. Dominant males, advantaged by their status, perform multiple copulations daily at the peak of estrus. As if watching fruit ripening on a tree, however, preferring the most succulent days of peak estrus, these males permit subordinate males to have early female access. Individual apes know each other quite well, with complex likes and dislikes. If a female desires to copulate with a particular male who is other than dominant, the pair can scurry off into the bush for a rapid clandestine copulation or, on occasion, separate from the band for a number of days in consort formation. Only when estrus is over does normal band comity return.

There has obviously been a dramatic change in the reproductive advance to the level of humans today. When might this have happened, and how might it have come about? What might be the program which could replace estrus?

No species could permit estrus to become eclipsed until an equivalent or yet more satisfactory program had been set into place. Estrus had to be superseded in order to be eliminated. For this to happen, certain new developments were needed, among them a less dangerous context for sexual activity to take place.

Early hominids were experiencing a new general ecology that involved ever greater distances of travel between feeding locations. Food sources became seasonal and were located ever further apart. Danger in travel from location to location was increasing, requiring greater protection by males. For an organism of her size, an early hominid female coming out of the rainforest canopy environment was uniquely vulnerable. Less than four feet tall, her gait was bent, shuffling, slow, and uncertain. While the swelling of her perineum during estrus advertised her vulnerability, her mental state at such times would have been less than composed. The social integration of the band at such times, partly agitated, was also to certain degrees disorganized. The need for estrus to be replaced became nothing less than evolutionarily imperative.

A special location for copulatory activity to take place came to develop as bands congregated at feeding locations after travel. Here safety became established by mere concentration of numbers. At the same time, these were occasions not only of relief but of jubilation. Chimps are well known to become exuberantly noisy when happily devouring the ripened fruit of a new found, well laden tree. Such feeding excursions no longer took place occasionally but were becoming a new way of life. If new copulations were to occur, such transient feeding locations would have been inviting, veritable oases.

Perhaps an interesting feature of human nature may have originated at this evolutional stage. Humans tend to desire to have intercourse upon any significant occasion of success. Due to its taking place in privacy, this is frequently underappreciated. However, when looked for, it can often be noted. No depiction of this emergent human need has been more popular than that of a sailor kissing a nurse in uniform in the heart of Times Square when victory was declared in World War II. Perhaps the most common occurrence of this emergent need occurs with the grand celebration which takes place when a cherished home team wins a dramatic game. It is as if human hormonal chemistry suddenly becomes reprogrammed by fresh relief from struggle and stress to success and joy, inviting a sexual expression. Perhaps as estrus began to fade hominid ancestors feasting under well laden fruit trees after arduous journeys came to experience such invitations to sexual arousal.

One feature of an advance to the experience of feeding locations would have been an improvement over the wild and little controlled sexual behaviors of estrus. Such locations would present new levels of safety for copulation, as well as occasions for initiating the emergence of voluntary controls.

If loss of estrus was urgent, and if possible locations for sexual activity were arising, whether or not a woman was willing to have sexual relations, would her genital tissues have tolerated such new activity?

As estrus proceeds in an ordinary monthly cycle, estrogen levels supporting vaginal access first rise, maintain high receptivity for some three to five days or more, and then subsequently decline. There exists appreciable variation – alike, with development, age, and health, with both individual and species differences.

One may be permitted, perhaps, to speculate that the estrus that came to be eclipsed in hominids would have approximated that of a modern chimpanzee or bonobo, our closest evolutionary cousins; perhaps longer than the modern chimp, who seems to have compromised reproduction somewhat for improved ground adaptation, and less than the bonobo, in whom sexual physiology would appear to have elaborated certain reproductively related processes in contexts of favorable arboreal adeptness, beyond the likely state that would have pertained for a common ancestor.

The course of this rise and fall generally follows the levels of estrogen circulating in the blood. The most intense copulatory activity tends to take place during peak days, with less frequent copulation on the one or two days before or after.

Appreciable evidence supports that an increase in levels of estrogen took place at this evolutionary juncture. At first, such a development might seem curious in a creature already intensely driven sexually. However, the new development, apparently a general thickening of the skin which could have emerged for thermodynamic reasons, appears to have had an opposite effect. It is well known that one of the actions of estrogen in the maturing young woman of today is a general doubling of the thickness of the skin. Although all of such an evolutionary increase may not have taken place at this time, even partially increased skin thickness may have been enough to produce certain changes. A particular case for a significant skin increase at this time can be made, however, since new thermoregulatory needs would also have been commencing to emerge in this major environmental transition. New demands were becoming placed on the skin for heat conservation and heat radiation in thermally more variable environments.

A small increase in the thickness of the vulva under increased estrogenic action would have had several major consequences. First, the ballooning of estrus would become reduced, with less reddening. With less swelling between their thighs, females would have been able to ambulate more easily, becoming somewhat safer in the presence of danger. At the same time, increased estrogenic actions upon those vaginal tissues which were only marginally functional would tend to advance these to full sexual availability. In fairly short order, by a simple conversion of perhaps two days before and two after, a four day period of vaginal access could become fairly easily extended to eight days. In time, further increases in estrogen levels could extend availability to the entire month. Such an elevation to increased breadth of normal tissue use would have been appreciably aided by increased vulvovaginal secretions, which would also be brought into action by increased estrogen levels. At the same time, with decreases of swelling it is likely that decreases of estral intensity and lessened behavior dependent upon cycling would follow.

An emergent advantage of increased estrogenic action at this point would have been that females could offer males sexual access over increasing numbers of days in the monthly cycle, and eventually over the entire monthly cycle. At the end of a trip for food, a sense of relief would be experienced by males and females alike. Females would particularly want to express appreciation to accompanying males.

Under such conditions, emotional actions could also have helped appreciably. Perhaps the major way in which emotional processes could begin to help would have been by increasing vulvovaginal secretions. This would involve both the extent of secretions in response to emotional arousal and increase in the number of accessible days. Vulvovaginal secretions serve as the normal preliminary action for all intromission. It is not at all unusual for such action under emotional stimulation to be copious. Many a young woman commencing her active sexual life today becomes so concerned about the profuseness of her vulvovaginal secretions under emotional excitement that she consults a physician, wanting to be reassured that she is all right. By rendering her days of marginal organ function more accessible as the changes of the menstrual cycle wax and wane, autonomic nervous action could thus serve to extend a the days of a female's sexual receptivity.

Upon noting the marked swelling of a female ape's genital organs during estrus and her marked drivenness, a first interpretation by a human observer tends to be that her state is one of intense sexual pleasure. However, the nature of female drivenness during estrus appears to need better understanding. An assumption of intense underlying pleasuring would appear to be only partially valid. Certainly significant pleasure normally accompanies any copulatory activity. Nevertheless, episodes of cutaneous pleasuring alone appear to be inadequate to explain the female's extraordinarily prolonged state of arousal at the great ape level. How and why could the later sense of pleasure in humans ever have replaced such an earlier, more intense drive?

It appears possible that two physiological programs of vulvovaginal arousal are here involved, with different types of "pleasure". Although genital pleasure would have been inherently involved, several considerations suggest that the drivenness of the female ape is more a state produced by intense pruritus than one of simple pleasure. This acutely driven state is only partially relieved by copulation, a feature which leads to repeated copulation with little rest over a period of days. Such behavior occurs even with tissues frayed. Her drivenness waxes and wanes in general accord with her swelling. This sustained copulatory program is more accommodative functionally to the needs of the band than would seem to occur from simple pleasuring alone, which would normally fatigue.

This functional program is suggestive of sustained pruritus as a result of acute swelling, involving histamine or histamine-like actions. Histamine is released strongly by any irritation of the skin. As the perineal region swells due to increased vascularity produced by the elevated estrogen levels of oncoming estrus, the skin correspondingly thins, a process that continues until it has become almost paper thin. As it thins, levels of histamine release steadily increase from tissue irritation. At full festooning, the skin has become so thin that it is flagrant red in color, warm, and fragile. Nevertheless, the thin skin does not advance to states of necrosis. The marked vascularity of this special organ state at this time is highly protective. Not only is viability preserved but any skin injury results in remarkably rapid healing.

A first role of histamine release is the conscious awareness of the irritation, with an attendant desire to correct it. In the bonobo or chimpanzee, the mental image likely to appear would be that of the male penis. As an immature female she would have observed older females copulating, while as a mature adult she would have experienced copulation herself. For such relief she approaches one male after another. If a particular male should not be interested in copulating, a great ape female can gentle the male's penis until he follows her wishes. At the same time, under the effects of histamine her mental state becomes more alert and active.

The acute endocrine arousal of estrus, as that of human sexual arousal, involves the release of endorphins. These temporarily decrease sensitivity to pain. The logic of endorphin action is that the temporary obscuring of pain makes possible the completion of essential behaviors, following which necessary repairs can be performed. Thus, whatever the state of injury that may be immediately present from heavy copulatory action, pain will not enter awareness until after this period of hyperactivity has passed.

The features of a possible transitional reproductive program eclipsing estrus suggested by the above involve a major modification of sexual arousal in male and female. As the genital tissues of female hominids further elaborate under estrogenic action, skin sensitivity and the use of tactile sensibilities appear to increase. Whereas emotional advances appear likely to become first called into action, as males and females learn to become more physically intimate tactile advances secondarily may have followed. Great ape females, already highly expressive creatures, would engage in more active touching through hugs, patting, caressing, stroking and some kissing. Arms about one another's shoulders are common. When so needed, the stroking of male genitals proves to be rarely unsuccessful. Such a program of increased emotional and tactile arousal would have high selective value.

Forces of natural selection would have acted to favor such a program of vulvovaginal change. Upon leaving the rainforest canopy, those females with the greatest perineal festooning, as sails set to the breeze, advertising the most and the most handicapped in walking, would tend to prove the most susceptible to predators. On the other hand, those with the most advanced estrogen increases, experiencing the greatest reduction of swelling, would have become the most advantaged. These and other considerations suggest that the loss of estrus would likely have commenced fairly early in the descent from rainforest habitats and would likely also have taken place with evolutional rapidity.

In the eclipse of estrus evolution appears not to have resorted to regression but, rather, to have produced major advances. Estrus became not so much "hidden" as replaced by permanent vaginal receptivity. In this transition it was not so much that ovulation came to be "obscured" as that a broader range of copulation became desirable for an enlarged scope of reproductive function, within which the usual days of ovulation would continue. Since a human egg usually remains viable for several days after expulsion from the ovary, fertility would continue, or perhaps become improved, by increasing the days of vaginal access.

When presumed histaminic actions are removed from the above sexual motivational equation, the remaining framework of sexual expression attributable to estrogenic action suggests a support of sensory genital pleasure that remains evolutionally but at yet nascent levels. The role of estrogens at early hominid levels appears to have been only a fraction of that later realized at human levels.

Although the sexual initiative by the woodland stage moved to the male, by later human standards sexual activity itself would probably have remained at no more than modest levels. With entrance into the new ecology requiring wider daily ambulation, it seems possible that some early advancements in androgenic actions may have taken place. However, androgenic hormone levels remain relatively low, and copulation remains still brief. While estimating measures of pleasure in sexual engagement must remain entirely problematic, the durations of intromission for our evolutional cousins and closest relatives are well known. Chimpanzee intromission, probably the closest to a last common ancestor, averages twelve seconds. Bonobo intromission averages fifteen seconds. In comparison, human intromission, realizing the new capacities made possible by the advent of privacy in sexual relations, averages fifteen minutes, and can extend much longer when desired.

The above considerations suggest that the basic framework of all subsequent hominid and human reproductive organization came to be set into place by changes early in the hominid career. Sexual drivenness in the female came to be replaced by personal willingness and desire. Cyclic and reflexive behavioral patterns were replaced by processes that were beginning to be cognitively and emotionally appropriate. Female initiation of copulation was replaced by characteristic male initiation. In exchange for the loss of a handful of intensely active days, sexual receptivity, setting the stage for future evolution, broadened to involve the entire month. Brief sporadic copulations would continue, until total ground safety could emerge. Emotional involvement advanced to become more than reflexive but less than constant. Pleasure remained largely sensory. In time, higher development of the central nervous system would elaborate and refine the relatively simple emotional and cognitive features accompanying sexual engagement at this formative level.

It is widely accepted that evolution takes place in brief pulses responding to acute environmental change, followed by lengthy periods of quiet consolidation. In the transition from rainforest to woodland, the basic simian reproductive plan appears to have been first replaced by an incipient hominid program that formed with the overcoming of estrus. When this program, sufficiently stabilized, became challenged at a later date by new environmental instabilities, a distinctly human reproductive program emerged. In this, the working hominid triad of safety, emotional desire, and tactile need served as the foundation for further human reproductive advances.

PART IV: AFTER
SEXUAL PRIVACY

THE HUMAN TENURE

Dramatic recurrent environmental swings produced
the ecologic context within which human evolution took place. Accommodation to these new contexts molded a reproductive ecology which was both
uniquely advanced and specifically human.

*C*ommencing 3.2 million years ago, polar ice caps began to form.
These trapped atmospheric water, cooling Earth's land masses and
rendering them more arid. In some areas deserts formed and progressively expanded. As the ice caps later melted, the atmosphere
changed to humid and warm. Of particular relevance to human origins, the climate of northern Africa, after forming large new deserts,
followed to experience climates of torrid rains producing enormous
lakes and dense forests. We are living today in a recurrently warm
interglacial period.

Increases in brain size that took place from the level of an australopithecine to the modern human offer approximations for estimating
the relative importance of each of these stages for human nature today. Although the fossil record has been greatly clarified during the
past decade, many problems of classification remain unresolved.
Many figures can be presented only as estimates. Whereas the australopithecine advance in brain capacity had progressed from approximately 375 cc to 450 cc, that at the level of the habilines is
estimated to have progressed from approximately 500 cc to 700
cc. The erectine advance progressed from approximately 750cc to

1,200cc, while that from an erectine to a modern human increased to approximately 1,350cc. In this course of development, it would appear that distinctly human nature today was primarily molded by the ancestral erectine experience. Over a 1.8 million year interval this experience enlarged cerebral capacities approximately 450cc. Exactness in relation to the status of the habilines has proven so far to be highly elusive. Habiline developments that appear to have increased human brain capacities initially by 300 cc or more in less than a million years, may be anticipated to have been particularly formative for emerging human reproductive nature, while developments at the recent level of *Homo sapiens* may be expected to present the most refined human developments.

The extended periods of drought which commenced during the late habiline and early erectine tenures served to modify the reproductive ecology of evolving humans dramatically. Repeated droughts of increasing severity produced periods of recurrently diminished availability of plant foods. For ages, for millions upon millions of years, fruits and vegetables had served as the main staples of the hominid and earlier simian diets. Increasing reliance upon meat now became necessary. This, however, involved a fundamental change in the ecology of human food acquisition. Although much is known and knowledge of this transition is improving, a large gap still exists toward clarifying this evolutional transition. Any interpretation attempting to envision the second major hominin reproductive transition can only aspire to be provisionally representative.

It has become well affirmed by ample paleoanthropologic evidence that males at this evolutional juncture took to scavenging. Much of the scavenging ecology of this time is understood. By watching for vultures circling over kills, the locations of game could often be identified, particularly a large animal such as a kudu or an aging rhinoceros recently brought down by one or more large carnivores. As is commonly noted in great apes today, hunting on a small scale, like the finding of a duiker immobilized in swamp mud or a monkey caught in an isolated tree, would be assumed to have continued in early man.

The experience of scavenging proved to be highly dangerous and physically demanding. Moving in small groups to potential locations of food, ancestral humans had first to chase away any competitors, perhaps hyenas or wild dogs who were feeding upon carcass remains. Though hungry, these ancestors as they approached felled game would often have had to decide whether to defer possession to any larger carnivore, perhaps a lion or a leopard which might be still feeding, or to drive it away, at least for sufficient time to sever a satisfactory amount of meat. Any desperation in their own hunger would have to be weighed against the estimated state of hunger or satiety of any feeding beast. Vultures were always a nuisance, but could be chased away readily.

Having once gained access to the carcass, after cutting through its tough hide, these early men would proceed to sever fleshy limbs through joint detachment. While some members of the group were severing meat, one or two others would be watching for the return of any chased competitors awaiting the human's departure or new scavengers drawn to the kill. Once the meat was obtained, these early men carried their booty to home bases or to safe feeding locations. There they would proceed to slice the meat on the bones, sometimes leaving cut marks still visible on fossils today, and to distribute slices of meat to hungry hands and mouths. Cooking had not yet been discovered.

Once a significant cache of meat had become secured, however, its processing in human hands advanced to realize a new order of mastery. The emergence of stone tool crafting at this juncture proved to be extremely fortuitous. The invention of stone knapping, a feat achieved by certain advanced australopithecines and developed by habilines, now proceeded to yield stone flakes with edges sharp enough to cut through tough hides and to slice meat. The sharp edges of the residual cores could be used as more substantial cutting instruments for severing joint attachments. These sharp edges were attained by flaking certain of the round, handy stones which had been used previously for hurling at hostile intruders and for cracking open nuts in dry season. The blunt ends of choppers could be applied to crush bone, yielding its highly nutritious marrow. While the invention and application of simple stone tools gave at least one line access to a new realm of food, several other hominid lines that lacked such access appear to have become extinct.

The persistent challenge to find adequate animal food placed heavy demands upon human acumen and repeatedly tested human endurance. During difficult droughts, hours of scouting, dogged travel, and careful tracking would often have been necessary to acquire sufficient food. From modern hunters and gatherers it is appreciated that not every excursion for food on the plains proves to be successful. It would have become impossible at this evolutional stage to have too great a hunting capability, for, whatever level of competence might be attained, challenges on a yet larger scale would always have remained. States of relative competence, improving by slow increments, pertained for the human condition throughout most of its two million year career.

The physiological advances which supported the male's distal ventures primarily involved wide ranges of androgenic action. Androgens at once prepare a man for more powerful striped muscular exertion, and, elicited by this action itself, serve to sustain the activity in progress up to completion, even masking for a while certain levels of fatigue. Androgens strengthen bone metabolism and increase the thickness of the skin over the entire body, as well as the ruggedness of subcutaneous tissues. The red blood cell count enlarges as the volume of the circulating blood mass becomes increased, improving body oxygenation. Androgens elevate body metabolism and sharpen mental acuity. Androgenic action contributes to ready, sustained elevations of blood pressure and to reduced reaction times, rendering sudden and heavy exertions able to become more readily accomplished. In sum, androgenic activities in the male serve to facilitate more effective male function in broader environmental ventures.

An effect of androgen in the male complementary to improving hunting prowess proves to be an increase in sexual need. As previously noted, if androgenic muscle impulsiveness presses him outward onto the plains early in the day, sexual needs arising toward the end of the day return him home. As scavenging and hunting increased to become regular activities in a new pattern of life, sexual needs would appear to have advanced from being intermittent and occasional experiences to ones of potential daily maintenance. Such programming may be underappreciated by modern humans living sedentary life styles, in whose daily activities androgens tend to be less fully elicited. Although the copulative action may well have been brief and initially less intense than that of modern experience, its role in hunting ecology tends to direct primarily toward a reward of sexual intercourse upon home return.

At the same time, similarly, woman's increasing ambulatory role in the environment as the hunter-gatherer way of life got underway in the face of frequently more severe scarcity would also have produced in her certain elevations of androgen production, though at far lower levels than in the male. Women not only foraged for food but had to utilize the nearby environment for the acquisition of water fetched in containers, building materials, and at later times for daily firewood, clay and other materials. As local resources came to be depleted, campsites had to be relocated, and as game migrated, women and children along with the men followed the game.

Adult male and female reproductive hormonal levels genetically programmed today appear broadly to reflect the historic ancient roles of the human genders in expanded environments. It is estimated that in such a traditional hunter-gatherer society as the Kalahari San of today the area worked by a woman tends roughly to approximate a twentieth that of a man. The genetic endowment of a modern woman places the level of her androgens at a tenth to a twentieth that of a male.

As was anticipated earlier in the reproductive advance which took place in the transition from rainforest canopy to woodland, the most promising course for incorporating parental involvement by the human male would have been through elaborating emotional processes. By definition, emotions comprise accessory energy systems which extend and differentiate increased activity in vital functions. Increased vaginal activity, which would now serve to produce deeper and more extended pleasure, could in time reach levels sufficient to produce emotional binding. Whereas the first hominin reproductive transition had realized changes to male initiative and to increased coital frequency, the duration of copulation at this earlier evolutional stage had remained brief. Now, with ground safety attained, the duration of intercourse became capable of indefinite extension and thus of producing emotional processes of more than passing duration.

In response to an array of new challenges relating mainly to attaining food, the human brain commenced to evolve rapidly. Although fossils revealing details of this transition remain yet to be discovered, evidently within less than a million years this organ, as above noted, enlarged in size from approximately 400cc to approximately 700cc. This near doubling of brain capacity would appear to have taken place at a rate which to this day has seldom, if ever, been matched. Such an extraordinary brain increase appears to reflect intense environmental pressures for adaptation and accommodation.

It was not only that new cleverness and agility of all sorts were in demand. An enormous new world of animal habits, of skill for scavenging and for hunting small game presented itself, a realm of newly important knowledge which would take ages to master. Although it is not clear specifically how, satisfactory communication between persons on food excursions became important. This took place perhaps by sounds, perhaps by gestures, and eventually by language. As men departed from others in the band who were left at safe locations for increasingly longer intervals, gradually a new working ecology was taking shape centering about home bases. With this, in time, different courses of maturation developed in evolving human males and females. New distinctly human gender roles steadily evolved.

A larger brain did not come without consequence. To prepare for the more complex adult roles needed by the new necessities for food in the form of meat, extended childhood development came to be required. Increased demands for parenting arose, alike to feed offspring through larger numbers of dependent years and to teach them more demanding environmental skills for adult roles. Whereas a mother at the australopithecine level might have had one developing child under her care at any given time, with the possibility of another about to reach maturity, by the level of a late habiline these responsibilities, by comparative life course history profile, would have enlarged to involve the care of a likely three. Such a burden for a mother attempting to parent alone proves to be unmanageable. For the satisfactory perpetuation of the band and for the larger species, the need for the male to share in parenting responsibilities with the female would have become essential.

The route to increased parental care which came to be realized by human males would appear thus to have coursed first through increased mutual vaginal activity, now becoming a more regular experience. At this stage, human evolution appears first to have commenced to advance vaginal pleasures to lengths and depths capable of producing emotional development of more than transient duration. Enlarging frontal lobe capacities which were evolving rapidly at this period would have made such emotional development possible. As earlier relatively undifferentiated emotional processes elaborated, more differentiated emotional processes at later and more integrated evolutional levels appear to have combined with processes commencing to individuate personality. In time, augmented human reproduction would come to involve the comprehensive synthesis of a well advanced sexuality, complex and often powerful emotional forces, and increasingly appreciative interpersonal involvement.

At the earliest human beginnings, sexual privacy had yet to be realized. When males would return from a scavenging venture, they would promptly distribute the meat in shared proportions to all adults, female and male alike, as they would expect other males to share with them in the future when they might be unsuccessful in the hunt. Although in a large catch all members of the band would receive some, not all portions were equal. Apportionment would tend to follow emotional ties, with sexually receptive females high on the list. Mothers would divide their portions with their children. Adults would still copulate openly, each copulative act being brief. As one male over time would probably copulate with numerous females, most females over time would probably copulate with many or even all males. Widespread copulation by females might prove to be more advantageous for the acquisition of meat from different returning males, some of whom were known to be better hunters than others.

The introduction of sexual privacy would appear to have changed this broad functional pattern, leading shortly to the establishment of a new, specifically human reproductive baseline. Through modestly elevated levels of estrogen sustained since the eclipse of estrus, vaginal integrity would be steadily maintained. Reflecting her heavy dependence upon the man, emotional needs in woman would likely have risen to newly developed heights. Emotional processes by this level would be fusing with individual personal appreciation, now increasing in each partner for the other. A new dimension of expanding interpersonal needs was developing, alike from shared parenting and from mutual sexual pleasuring. Increased levels of testosterone in the male resulting from wider environmental activity were rendering him to become a more persistent and a more regular sexual partner. With inherently programmed sexual inhibition still absent, with sexual pleasure yet direct and uncomplicated, and with copulative duration now somewhat lengthened, it appears likely that a relatively simple and well integrated state of human sexual function may have come into effect.

Particularly through new actions of testosterone relating to widening environmental action, the sexual experience, though still brief, was becoming one of shared desire and mutual appreciation. In that it was unencumbered by levels of inhibition yet to evolve, this experience was proving to be relatively simple, and in the sense that new appreciation outweighed any individual differences it was interpersonally relatively pure.

When all is well – when a woman's focused desire upon one man is strong, when love enfolds their copulative action, when summation has had sufficient time, and where there has been sufficient sexual learning – climax in woman can come with ease. However, the significant powers of inhibition which have evolved in her nature do not permit in her the same readiness for or ease of climax as in the male. It is possible that, in neurological compensation for the greater summative work which appears to be part of her nature, her climactic experience, when it occurs, may tend to be more personally fulfilling. The validity of such a possibility, difficult either to prove or to be disproved, must remain conjectural.

As the feature of sexual privacy worked out its first major consequences at the level of the late habilines and early erectines, the basic reproductive program of humans appears thus to have received its foundation. The integration of sexual, emotional, and interpersonal dimensions at this early human stage appears to have established a basic design both for millennia of further reproductive development and to which, at times, higher processes might have occasion to revert.

Is it too much to wonder, perhaps, whether such a foundational sexual program existing before human intensities and complexities had arisen may have left its mark in the present? When a couple have become sexually experienced, when they still have desires for sex but are no longer driven, when romance is respected but occasions seem not always immediately to call for it, when, perhaps tired or simply more mature, their energies for sex no longer abound, when a couple's inhibitions no longer inhibit and interpersonal complexities no longer constrain expression, is it too much to wonder whether the uncomplicated, direct, warmly affectionate sexual experience that may ensue is far different from that which appears to have formed at these earliest human beginnings?

Whatever may be the answers to such questions, the present study proceeds in support of the argument that sexual privacy was preconditional for much of subsequent personal human development. A new realm of shared experience was emerging, unseen by the naked eye and quite unanticipated by the evolutionary course of developing external powers. Sexual privacy appears to have served now as a focal component for a new inner domain of evolving human reproductive nature.

VIII

TRIPLY TIERED HUMAN SEXUAL DYNAMICS

Human sexual nature has reptilian roots, mammalian stock and human flowering.

*F*undamentally, the integrated neural dynamics of human sexual nature is three tiered. This reflects the fact that, as neurophysiologists often emphasize, the human brain is a composite of three brains which have successively evolved. More than half of the mass of the brain, the cortex, is specifically human. This engulfs a much smaller mammalian brain, the limbic system, which, in turn, engulfs a yet much smaller reptilian brain, the thalamus. Although the actions of each of the lower two strata normally are referred upward to the human cerebral cortex for behavioral expression, each level characteristically molds behavior according to the parameters appropriate to the evolutional stage which it represents. Thus the reptilian brain mediates actions largely of stimulus-response nature, of short range and tending to be of brief temporal duration. The main reptilian reproductive inheritance is essentially the basic sequence of behaviors which constitutes sexual intercourse, the foundation of the human program. Upon this, the mammalian brain weaves emotional processes, particularly those higher emotions which arose in mammals. These involve a dozen or more basic emotional patterns such as joy, friendliness, disgust, and pleasure, which are distinguishable in the great apes but, most of all, the warm affiliative feelings which we describe as love. These more elemental emotions contrast with cortical emotional expression which tends to be emotionally refined, subtle, more differentiated, and cognitively far more widely engaged. In everyday experience the normal workings of these strata, all summating together through cortical integration, are felt respectively as sensory need, emotional need and needs for interpersonal appropriateness in general sexual expression. It is interesting that these three realms are commonly recognized in everyday experience as relatively distinct dimensions of human reproductive sensibilities. In women's magazine literature these are described respectively as lust, love, and the importance of a Mr. or Mrs. Right.

The triply tiered structure of the human brain reflects its three-staged evolutionary history. In the seas, the centralizing nervous systems in early vertebrates first developed specific nuclei for negotiating functions essential to the continuance of life, each at its relatively simple level. These vital centers regulated food acquisition, digestion, excretion, reproduction, body movement toward favorable stimuli and movement away from noxious or unfavorable stimuli.

As vertebrates moved from the seas onto land habitats, such nuclei for vital processes came to be organized into major functional systems. These first elaborated basic stimulus-response patterns through thalamic and hypothalamic advances, the reptilian stage. At the next level of major development, the mammalian stage, the organized vital systems functioning at the thalamic level became augmented by accessory energies, supplied according to their particular importance to the organism. Forming a paleocortex, these limbic level systems comprise basic emotional elaborations. At the level of the neocortex, the realm of the human advance, vital processes had become so proficiently managed and accessory systems so effective that only a refined awareness of inner vital perpetuation was needed. Primary cognizance could normally be given to external environmental events, permitting rational action. In such higher states, emotions may be present as fine background feeling or they may be absent. Should inner vital function sufficiently weaken, however, such a deprived need tends promptly to exercise control over more elaborate, higher processes.

Although all three tiers normally act together, their proportionate contributions can vary markedly from time to time, person to person, and situation to situation. This reflects the fact that as each higher tier evolved, existing lower tiers were retained in case higher processing proved to be inadequate. Control is normally exerted by cerebrocortical levels through the exercise of metabolic rates higher than those pertaining in the lower levels. In essence, higher neural levels act by preemption of action. On the other hand, lower levels can influence or even exercise control of behavior through producing higher intensities of action capable of overriding higher level processes.

Although words are notoriously imprecise, the rewards of these three levels are often described differently. Those at the stimulus-response level comprise pleasures; those at the emotional level are more lasting satisfactions, while those at more inclusive personal and cognitive levels are often described as deep gratifications.

A fairly obvious example of how such combined integration functions may be noted in the spectrum of emotional responses relating to anger. In mild frustration the cortical processes focus upon the frustrating agency, working through normal energies or perhaps with mildly increased efforts. Comprising simple mental hostility, such modest action normally enjoys the full integration of the nervous system and uses its most differentiated capacities. Such actions involve increased effort without compromising higher function. Hostility processed through normal cerebrocortical action presents the functionally most integrated human response to frustration.

Should this effort at resolution fail, however, increased efforts to meet the frustrating agency become needed. Neural action then begins to rise to stronger degrees of emotional engagement mediated by higher levels of limbic work. Employing cruder and less economic neurological processes in functional compromise, these sacrifice certain highly differentiated and highly integrated processes. In this state simple hostility in response to frustration proceeds to become frank anger. As exemplified by an often reddened face, the mobilizations of anger involve an elevation of body metabolism itself. Anger utilizing varied emotional powers for augmented response presents the basic mammalian program of response to higher frustration. Impulsive, poorly controlled hostility is commonly noted in apes.

Should the functional compromise negotiated by ordinary anger fail to overcome the obstacle, in turn, yet more elevated levels of neural activation take place. Anger gives way to rage, a state again less integrated, less differentiated and less economic but more powerful in its response. Whereas anger involves the eclipsing of such immediately unnecessary functions as growth and reproduction, a significant internal reorganization and heavy functional compromise, rage involves so much sacrifice of the organism's normal capacities and metabolic costs may become so heavy that the system places itself in jeopardy from deficiencies of normal integration and internal balance. These losses occur in spite of the fact that rage elevates body metabolism to levels higher than anger. Wild rage can also be observed in apes sufficiently frustrated.

It is no surprise that sustained anger and rage not only predispose to poor judgment and interpersonal callousness but to numerous disease states. In humans coronary artery disease, hypertension and personality disorders are well known to have chronic anger as contributory agencies.

The normal human sexual experience involves the combined actions of all three phylogenetic tiers in differing proportions at different times. Reflecting its recent, specifically human evolutionary environmental context, the major role of the cortex is one of interpersonal involvements and diversity of voluntary action. This both realizes the highest definition of a human personality and matches personalities with the finest sensibilities. The role of the cortex is circumspect in the widest sense. Deeper needs which are emotional and sensory are present but at the initial period of sexual engagement, which can be prolonged even for years, remain in the background until appropriate conditions have been met.

Human cortical action involves first the external realm. Each person must find the right person for intercourse with the right emotional fit and appropriate societal qualifications. In particular, the sexual partner must be the right person. Even when casually experienced, human sexual expression has difficulty avoiding being inescapably personal. The situation must be right, not only secure but separate from other persons, and comfortable. The time must be right; providing at least a reasonable temporal interlude, and preferably a luxury of time. The participating person must feel right; not only having sufficient desire but also being in sufficient health. The social and cultural context must be acceptable; even features of different class, upbringing or religious beliefs can impede or prevent the favorable onset of sexual relations. And, as almost every young person well knows, such a listing is but partial. These are all problems in the outer world which require to be negotiated by the specifically human brain before the inner domain can proceed to meet its sexual desires.

Once a pair has attained appropriate external environmental preparation for intercourse, emotional and sensory needs commence to exert more immediate influence, working through the emergent prism of privacy. The apportionments of these levels, however, continue to vary so widely according to person and circumstance as to render the highest respect for neurological subtlety and appropriateness.

A human experience of intercourse that progresses to orgasm demonstrates dramatically the combined interactions of all three tiers. These successively phase into one another. Although from individual to individual the particulars of the experience vary widely, in its basic progression the program's course involves a pattern of successive reversion to functional dominance by less complex, more intensely active earlier evolutional levels.

As the interest and duration of sexual desire increase, neural integration proceeds to follow successively more intensely dominant neurological states. These augmented states are first limbic, emotional. As the state of arousal continues to increase, limbic mobilizations come to engage thalamic processes more heavily. In this process, neocortical correlates become eclipsed early. As the contribution of emotional centers intensifies, these fuse more and more with intensifications of thalamically-based expressions of sexual need itself. As climax approaches, the intensely elevated levels of need tend to obliterate not only all rational processes but also all emotion except basic excitement itself, the primal undifferentiated emotional state. As orgasmic climax proceeds to become inevitable, reflexive sensory systems assume complete control.

Upon completion of the experience of orgasm, neural regnance promptly returns to the normal rational world. The participants, however, are no longer the same. The experience has left its record in emotional memory and in shared appreciation. In early adulthood the male experiences more intense sexual need, with less broadly developed emotional needs. The young woman, on the other hand, has more intense emotional needs at this time and is less driven sexually. This complementary relationship tends to prevail up to the fourth decade. As the third decade wanes, the physical powers of the male, including his sexual drive, begin their long age-related decline. Although his sexual drive lessens, his emotional needs tend now to increase, a pattern which may well progress for decades. If the man's partner has become well experienced sexually, she may attain the height of her sexual powers in her fourth decade, with her emotional needs at this time tending to become perhaps less intense. In later decades, although sexual needs continue to be present, emotional needs in both sexes tend to predominate.

The role of emotions as accessory need systems here becomes quite evident. These serve to sustain adult reproductive capacity through the two decades which require maximal parental effort, an ideal fitting of capacities for work by parents for developing human young. By the sixth decade less intense parental work has usually become necessary.

The triple tiering of human sexual nature may show itself in myriad ways. Although the composition of the human nervous system is such that all three tiers are normally involved, variations in tier participation are wide. A middle age adult in whom sexual drive has lost intensity and in whom emotional needs are not strong may come to regard the experience, though necessary, to be much overrated. He may find himself wondering, after all, what the shouting is about. On the other hand, a young adult in early stages of copulative learning with as yet but limited powers of control may find the experience, though lovely beyond words, breathtakingly tantalizing and even at times addictive.

There usually exists an unspoken mental contract within which a sexual experience is undertaken. In a one night stand sexual needs tend to assume the predominant role.

Emotional factors are usually set aside; the partner is mainly a person who is sexually available. Although society, appreciating the marked vicissitudes of sexual need in different persons at different times, tends to be tolerant of such liaisons, it should be no surprise that, lacking emotional involvement, such couplings fail to produce bonding. Indeed, they are often chosen for this reason. The particular danger of the one-night stand, however, is that almost by stealth, at times, it can result in personal or interpersonal injury. As elsewhere, human sexual nature may here reveal its inherent triple tiering. Subtle burnouts may occur, or subtle resentments accumulate. Although personal involvement may seem to be irrelevant, it is difficult for human nature, particularly over time, so to regard it. Other personality variables may be able to protect the psyche, but the neglect of personal considerations and of emotional involvement in casual or one-night stands can take a toll on the person.

The obverse is also true. The glory of the triply tiered synthesis occurs in those realizations in which the lower tiers contribute generously to the higher, specifically human brain, enabling it to play its widest role: when personal appreciation is deepest, love strongest, health most hearty, and the variables of the external context most supportive. This can take place only after the above noted necessary human external prerequisites have been sufficiently met. At the human level, both the sexual needs of lower levels of vertebrate evolution and the emotional levels of prehuman ancestors have become markedly enlarged. As a great ape is a more sexual creature than a reptile, a human is a creature far more sexual and more finely emotional than a great ape. It is the nature of the cerebral cortex to integrate these forces into the entirety of ongoing experience.

The distinction between love at a more basic mammalian level and love fully expressed at the human level is described by Jane Goodall. From her observations of chimpanzee life at the Gombe Reserve she notes that these genetic cousins, so close to human ancestors, show a basic love but no refinement, mere shadowy forerunners of what was to emerge in humans. Human love in its full expression, the product of broad frontal lobe action, is much more. As she describes it:

> I cannot conceive of chimpanzees developing emotions, one for the other, comparable to the tenderness, tolerance, and spiritual exhilaration that are the hallmarks of human love in its deepest and truest sense. Chimpanzees usually show a lack of consideration for each other's feelings… For the male and female chimpanzee there can be no exquisite awareness of each other's body – let alone the other's mind.

Amidst the varying circumstances of life, should a woman engage in sex with a man whom she does not love and then, subsequently, with a man whom she does love, she will report that the differences of both sexual pleasure and inner gratification in the latter experience in relation to the former are beyond comparison. As sensibilities of pleasuring and emotional appreciation fuse with maturing, many a man comes to find the most exciting part of the sexual experience to be not his own climax but the woman's satisfaction from orgasm.

How do these three tiers contribute to produce sexual maturity? Each tier by itself can prove at times to be so strong as quite to overwhelm the person.

The temporal dimension of each tier reflects its evolutionary origin. Although sexual pleasuring can be incomparably intense, alone its effects can be so short acting as to be gone by the next day. Emotional satisfactions characteristically last longer. The reverberations of loving may enhance daily life for weeks or for years, even when interpersonally unworkable. But life is learning, and emotional experiences do not always last. Gratifications from interpersonal involvement last the longest of all. The tier compositions capable of producing the strongest bonding and the most promising long range growth combine emotional enjoyment with interpersonal compatibility.

It is often possible for a person to gain insight concerning the triply tiered composition of any particular experience of intercourse. On the morning after, or an equivalent time permitting restored objectivity, one may ask: How deep were the sensory needs? How much were emotional needs involved? How much interpersonal affirmation took place? In any person, different experiences give different component summations at different times.

IX

THE EMERGENCE OF HUMAN GENDER ROLES

The human male is primarily masculine and secondarily feminine; the human female is primarily feminine and secondarily masculine; together, they make optimal use of any given environment.

A major advance which takes place at any evolutional level is likely to have been anteceded by minor advances at an earlier level. The warm-bloodedness characteristic of the mammals was preceded by its apparently independent development at the level of the reptiles in certain later dinosaurs.

Human masculinity and femininity appear to have had relatively minor antecedent functions at the australopithecine level. If a traveling hominid group were to have spotted an antelope nearby, particularly one young or in any way injured, one or more males would likely have separated themselves from the group in pursuit. Similarly, when a chimpanzee group today senses a threat from a nearby leopard, leaving females and children behind, males gang together, moving cautiously toward it in order to chase it away using scare tactics, stones, and sticks. Occasions of male separation from the rest of the band for food or safety are common among primates.

The establishment of home bases over a yet unknown interval of time at the outset of the human career served to introduce what was to become a characteristic human division of labor by gender. While males had to travel ever farther to capture game in ever more severe dry seasons, women, children and the elderly were left behind in a consolidated group, usually located near water in relative safety. Earlier in evolution as males made forays into the near environment women and children might have been left at convenient safe locations for only minutes or hours. Later, however, as male groups took to scavenging and thence to large scale tracking and hunting, periods of separation would have lengthened, not uncommonly for one or several days.

Human evolution seems to have accommodated the divergent environmental activities of men and women by developing gender physiologies. Women came to specialize in carrying, alike nursing infants and gathered foods. Through a gait made smoothly horizontal by hip mobility, they became able to carry water in containers on their heads without spilling, or firewood in bundles. Males came to specialize, first in scavenging and then in hunting over widening ranges of terrain.

Gradually, over a period of one and a half to two million years, certain physical and mental characteristics developed in evolving human males. Before youthful males could venture afield to participate in hunting , their voices had to turn, an action of testosterone enlarging the larynx. Male voices needed to be of lower pitch, close to the   lowing of contented animals, lest they produce alarm.

These facilitated more extended environmental roles. They were mediated by diverse advances in a cluster of hormones which included testosterone, adrenal hormones and growth hormone. As a result, human males became taller and stronger. As a man's voluntary musculature enlarged, his bones also lengthened and became stronger. Male endurance markedly increased, while at the same time motor skills from walking to hurling objects became more refined. In time, a man's upper body powers developed to become twice that which developed in a woman.

Testosterone has major actions favoring environmental mobility and prowess. Testosterone increases the production of proteins over the body diffusely, not only in muscle and bone but also in skin and other body organs, thereby developing a larger, more powerful, and more physically resilient person. Testosterone raises body metabolism normally from 5% to 10%, and even at crucial times to as much as 15%, producing a behaviorally more powerful individual. At the same time, as if preparing for possible severe injury, testosterone increases the red blood cell mass up to 20%, as well as enlarging the total body fluid volume in preparation for possible loss

Along with these developments certain personality characteristics evolved. Testosterone, in particular, elevates body metabolism, causing the male to become behaviorally more active. A cluster of features developed which correlated well with each other and are still found in modern personality inventories. As revealed in such tests, these include such characteristics as assertiveness, abilities to be dominant, competitive, forceful, aggressive, ambitious, athletic, independent, individualistic, self-reliant and self-sufficient. Masculine capacities include the ability to act as a leader, to take a stand, to take risks, to defend beliefs, to show strong personality features, to think analytically and to make decisions easily.

Such a description of human masculinity clearly reflects the tracking of game on the plains, alone and in groups, over countless millennia. What is of interest is that the physiological advances, here but modestly profiled, appear entirely consistent with the mental attributes which developed during this same period and which persist with full validity in our advanced world today.

Individuals in familial groups, primarily women but also to certain extents children, developed analogous characteristics facilitating their roles in home-based activity. In women these were mediated by advances in estrogens, progestins and the lactogenic hormones. In children, home-based activities were facilitated by somatic and mental developments, none more evident than the slowing of growth during puberty before later renewed acceleration to adolescence and maturity. In both boys and girls this slowing of pubertal growth permits longer and fuller maturation at home base before the commencement of the adolescent growth spurt leading to maturity. A more consolidated base of home learning serves as foundation and basic reference in the role of departure and return of the hunter.

In females, estrogens terminate body growth approximately two years earlier than occurs in males. This produces shorter and smaller-boned adult women, prepared for reproduction at a slightly earlier age. In modern marriages, women still tend, on average, to be approximately two years younger. At home base women primarily tend to the young. Nine months of gestation is followed by one to three years of nursing, which can be longer when starvation threatens the group. As each child is weaned, woman in primal contexts prepares for another pregnancy. One result of this maternal programming is that she usually has several children developing under her care at any given time. Mortality rates in all-aboriginal settings are high. By any modern standard, maternal losses due to pregnancy and childbirth can be extremely high, while mortality rates for newborns within the first year of life in preindustrial cultures, even more devastating, can reach as high as 50%.

Women at home bases gather vegetable foods in the vicinity. These include fruits, tubers, roots, tender leaves, nuts and occasionally the treat of wild honey. Not all male hunts are successful, in which event woman's food supply supports the entire group. More often meats and vegetable foods are able to be mixed. Women maintain the fire and do the cooking. A constant supply of firewood and water is necessary, requiring almost daily excursions into the surrounding environment for new supplies. Women care for the sick, the aged and the disabled. They fashion clothing and weave. Not least, they teach language to children, and, with it, transmit the essentials of the given culture to the next generation.

Woman's developments in the familial realm are similarly reflected, alike in her physiology and in her personality characteristics. As noted in a modern personality inventory, Bem's feminine index includes being affectionate, cheerful, sensitive to the needs of others, gentle, compassionate, tender, sympathetic, warm, loyal, understanding, soft-spoken, shy, gullible, yielding, flatterable, eager to soothe hurt feelings, and being loving toward children.

To check the validity of the masculinity and femininity complexes discovered in her studies, Bem checked each for consistency, for shared internal correlation, and for independence, each complex from the other. Her neutral inventory, used as control, included such items as being adaptable, conceited, conscientious, conventional, friendly, happy, helpful, inefficient, likable, reliable, moody, sincere, secretive, tactful, truthful, theatrical, unpredictable, and systematic.

Undoubtedly, the features of each of these two complexes, as well as those of the neutral control group, could be enlarged. However, the essential fact here disclosed is clearly that masculinity and femininity are each distinct and unique dimensions within a human personality.

It has long been clear that every personality includes both masculine and feminine characteristics. It has also long been equally clear that these complexes contrast. This is well represented in their endocrine profiles. In human males, masculine hormones quite predominate over feminine, whereas in human females feminine hormones quite predominate over masculine. This hormonal contrast fundamentally reports that for untold thousands of years ancestral humans specialized in complementary major environmental roles, while specializing far less in respective minor environmental realms.

In particular, males came to develop advanced proficiencies for negotiating the great, outer world. These came to include more marked left-brained dominance and more advanced spatial capacities. A man's brain has been described by one neurophysiologist as a problem-solving machine. Females came to realize more powerful visceral capacities, ranging from meeting sexual needs to meeting simple body comforts. Women's verbal facility is usually greater than that of a man, while, due to much larger connections between left- and right-brain hemispheres, she tends to have a more comprehensive intelligence. At any given moment, she is more likely to know where various objects are, such as children, and what is happening, such as various foods in progress of being cooked. She enjoys greater finger dexterity, and has a better sense of smell. Whereas high levels of testosterone and a greater behavioral activity level produce a sexual drivenness in the male, woman's hormonal complex directs her nature to a more immediate sense of the perpetuation of the species. This centers about the family first, accepting the male as a father and provider, and being intensely loyal to him. A woman needs a satisfactory level of masculinity, alike to relate to a man and to his world. She also needs sufficient masculine features to fully express her sexual needs for successful mating. Not to be neglected, she needs a certain understanding of the wider world to satisfy her human mental nature, lacking which she may feel restricted.

The role of home-based activities for the male, though less evolutionally developed, proves to be no less essential. Lacking these, the human family fails and the species loses continuation. For insuring the perpetuation of these, as noted, the early, most highly formative years of development take place within the context of the home environment. Femininity in the adult male renders him capable of measures of personality integration with women and with children. With sufficient feminine features, he becomes able not only to maintain himself and his family within the cultural group but also becomes better able to cooperate with his male peers.

The basic relationship of these contrasting gender complexes, each with the other, is noted to be profoundly complementary. Masculinity and femininity function as entirely independent variables in an adult personality. A man or a woman may be either high or low in one or both. Under favorable circumstances human male and female personalities are able to match each other well. In broad ecologic perspective, this complementary pattern is typically primate. By this complementary division of labor any given environment, as extended as may be needed for the species, becomes maximally utilized, alike for food, safety and other needs.

Each gender by its role division becomes something of a mystery to the other. It is the wide interplay of the features of masculinity and femininity in the modern human male and that of femininity and masculinity in today's woman which act alike as delight and perplexity. The ancient complementary roles prevail as a usual basis for family formation and its continuance, with greatly enlarged margins of variability introduced by the many new lesser environments of civilization and by different cultures.

The specialization of daily labors by sex served to create in humans a new theater of shared concern and interest. It became the role of the man to explain and interpret to woman the nature of the large outer world, while it became the role of the woman to detail to the man the events of the home arena, particularly daily developments in the children. This realm of shared concern, novel to primate experience and specific to humans, became a source of pleasurable necessity which also served to bond parents together.

Male specialization in distal environmental activity for thousands upon thousands of years led to a differential use of vision by the sexes. Leaping from tree to tree in search of fruit and other foods, all apes prove to be highly visual creatures. In human males exploring new vistas, elaborations of testosterone combined with certain genetic advances to produce certain more advanced visual specialization. Whereas vision normally is the least emotionally endowed of the senses, upon seeing a possible catch vision proves able to elicit appreciable excitement. With the human male's prolonged absence from home, images of a sexual partner tend to emerge in his mind as the day wears on. These, similarly, elicit inner arousal, helping to return him home. Unlike the human female, whose arousal relies upon other factors, particularly emotional processes, the human male becomes sexually responsive by visual stimuli alone.

From the most ancient times, women have been highly aware of the male's visual sexual excitability. The earliest cave paintings consist of female images undoubtedly drawn by men, while the earliest artifacts of civilization include women's mirrors and cosmetic kits. In our own day we have Victoria's Secret and the vast cosmetic industry, well maintained by women. Many a woman in today's world would not be seen without eye liner. Indeed, every young woman knows from childhood on that a beautiful face proves to be an enormous asset in her relating to men, and in due time to one in particular

Visual excitability serves the male broadly as a cue and an aid to sexual arousal. The absence of comparable visual excitability in the female leads her to rely upon slower factors, such as emotional processes and touching for her arousal.

Favorable gender matches, however, though normative for humans, prove to be far from automatic in their realization. As cultures become more complex, highly favorable matches tend to become more difficult. What proved to be an easy match between a hunter and his wife presents as a far simpler order compared with that of the urban man and woman, each of whom may be engaged in different types of work lacking any inherent complementary nature through which one can do anything for the other.

In terms of matching, several general problems tend to emerge. As excessive masculinity in a male may render him unable to maintain a family, inadequate levels of masculinity, similarly, can render him unable to play an adequate fatherly role in a family setting. In woman, excessive femininity, a personality too limited, may render her less able to relate satisfactorily to males, while inadequate femininity may render her less interested in acting as a mother in a family context. She may opt for a life as a career working woman.

Appreciating the complexities of civilization today, it is no wonder that humans find the problem of a favorable matching, one traditionally central to the human lifetime, so often difficult. On the other hand, it is also no little wonder that these ancient gender proclivities persist so powerfully. Fortunately, broad leeway exists in these attributes and in their matching. Different individuals may find highly diverse solutions. A masculine woman may find most favorable to her makeup a feminine man, or a highly masculine man may somehow quite favorably and mysteriously match with a highly feminine woman. Highly important is the fact that these attributes are not fully formed at maturity but, to certain extents, also develop secondarily over the lifetime. This presents room for significant ongoing mutual accommodation. Mothers who have all male children tend to become more masculine, coming to understand men better over the years. Fathers who have all daughters tend to become more feminine over the years, appreciating women more.

Yet, this said, ideal gender matches ever tend most to follow the ancient profiles. This appears to be true in most cultures. The most common solution in civilization today proves to be for a man to work full time and for a woman to divide her work time out of home according to situation and need.

In all pairing, in all cultures, masculinity and femininity remain but two aspects of satisfactory personality matching. Other major dimensions of compatibility include family backgrounds, sibling relationships, cultural compatibility, similar interests, sexual compatibility, and even a shared sense of humor.

As they are realized today, traits of human masculinity and femininity are not highly specific variables of human nature but, rather, comprise broad complexes summating many features of personality which relate to mating compatibility. Many different combinations prove to be workable. Like the hormone profiles which they largely reflect, rather than being determining agencies they are directive processes within the personality. However, for the man to be more masculine than the woman, and for the woman to be more feminine than the man, appears in overall balance still to prove the most workable general relationship of all.

How do such generic definitions of masculinity and femininity fare with those of cultural relativism?

On the surface these do not always match. Comparative studies of different cultures reveal a human diversity of almost total proportions. What is masculine in one culture is not in another, while what is feminine in a given culture proves not to be in another. Gender roles in different cultures prove often to be completely opposite, one with another.

Historically, the masculinity and femininity of cultural relativism attained more complex definition as cultures evolved following the invention of agriculture. These reflect different environmental adaptations and different accumulations of cultural development. As long as cultures continued in hunting-gathering ways, generic definition prevailed.

Generic definition has not disappeared but, rather, has become merged with human capacities for cultural elaboration, with widely diverse results. Yet, in almost all cultures the lifting of heavy objects is done preferably by men, while in all cultures the care of young children and cooking tends to fall upon women. Elaborations of culture tend to diverge partially but not completely from the aboriginal hunter-gatherer base. In our own culture, which retains still much of the historic frontier where the division of labor by gender continued to fall largely along generic lines, clear currents of generic human masculinity and femininity remain quite in evidence.

Cultural "masculinity," what men do in one society, and "femininity," what women do in another, so appreciated, present elaborations of generic masculinity and femininity. Though with general similarities, these show remarkable capacities for human adaptation. With increased complexities there tend to come increased problems of gender compatibility. The ability of human nature to adapt often comes into question, with no satisfactory answer yet in evidence.

In an overall view, it remains impressive, even a matter of wonder, that in modern humans in advanced technological settings the masculine and feminine complexes which evolved as ancestors came to negotiate proximal and distal environments and which served them so well long ago should still remain relevant to human happiness today, long after the working frameworks of the environments which molded them are gone.

X

FROM SIMPLE SPASM TO DEEPENING ORGASM

> At an early date in the human tenure, the role of re-
> productive climax changed from serving to terminate
> copulation promptly to facilitating its duration. The
> fulfilling experience of human orgasm came to act as
> a stabilizing force in daily living, as well as a motiva-
> tional force toward personal maturation.

As an accompaniment of internal fertilization, the phenomenon of climax appears to have emerged in vertebrate nature at the level of the reptile. At the earlier level of the amphibians certain antecedents appear to be present As noted in the frog, fertilization is external, involving grasp reflexes that become coordinated with female abdominal contractions. As the female expresses her eggs, male sperm is expressed in a series of repeated combined efforts, realizing immediate fertilization. Although a certain amphibian sense of "pleasure" produced by the frog's simultaneously engaged reward system would be expected to accompany this action, no evidence of a climax is noted.

In the reptile, however, where fertilization has advanced to become internal, a more developed reproductive proficiency that concentrates sperm ejection into one short intromissive interval would be more likely to realize more distinct sensibilities of such behavior. In this view, through more developed reward system activity, the first vaginal contractions to act with summation upon inserted penal structures would likely have produced a first summative copulative awareness sensed as a climax.

At the formative level of reptiles, two terminal oviducts fuse distally to form a crude receiving chamber. The external os of this organ, like any bodily opening to the external world, comes to be ringed by particularly sensitive cells responding to touch and to pressure. As the male inserts his penis, or in some reptiles his paired hemipenises, into the primitive vagina, sensory cells responding to pressure and to touch upon deep penetration send messages to lumbar spinal cord centers. Here clusters of primed motor neurons send back reflexive messages to the vagina, producing prompt rhythmic contractions of its musculature. This contraction acts as an immediate stimulus upon the penis, evoking ejaculation. Though little or no more than a simple stimulus-response action, this sequence appears to have conscious representation. Upon its completion, the pairing reptiles decouple and, parting, go their separate ways. This lower vertebrate sequence presents the foundational program upon which all higher vertebrate reproductive physiology will build.

At the level of the simians, a cluster of sensory cells at the vaginal ring has elaborated to form a special neural structure, the clitoris. Positioned on its presenting margin, this organ is so placed as to respond only to full penile penetration. At the same time, having a concentration of neurons unsurpassed anywhere else on the surface of the body, it has become specialized to respond to pressure forcefully. Under clitoral action and through spinal processing, strong muscle contractions of the vaginal wall engage the fully inserted penis, evoking ejaculation. This organ is so sensitive that even small direct pressure upon it can be sensed as excessive, a feature that tends to hold the penis in place while ejaculating. Almost immediately upon the ending of her spasms the female withdraws her entire body. Thus, as first noted in the reptile, the evolutionary role of the clitoris appears to have been an assurance of fertilization through mandating penile depth, involving as brief an interval of time as possible.

The reason for this brevity, as often noted in this study, is that under states of pleasure, which may increase vulnerability, the female terminates copulation as rapidly as possible as a necessary safety precaution. Animals in general are never more susceptible to predation than when mating. Every youngster who has gone crabbing knows that these crustaceans are never easier to catch than when doubling, in which state they can practically be taken by hand. Similarly, when mating, a dog keeps a watchful eye upon the surroundings. The female withdrawal reflex, which occurs universally with deep penetration, minimizes coital timing and temporary organismic vulnerability.

At the human level the basic scenario of vertebrate copulation abruptly changes. The emergence of sexual privacy makes this transition possible. With intercourse no longer required to be brief, the duration of copulation is able to proceed open-endedly, limited no more by time but only by such variables as health, pleasure, desire, and convenience.

The advance from simian to human reproductive programs involved significant organ changes. In males, penile size in the erect state increased from an approximate four or five inches at the level of the chimpanzee to an approximate six to eight in a human, with penile girth also multiplying numerous times. In addition, penile sensitivity in the erect state markedly increased.

In women, the human advances of clitoral function appear to be no less marked. With human sexuality advancing significantly, it might be expected that the clitoris would also enlarge. However, it does not do as expected. Rather, it elaborates at human levels by a diminishment of size. Whereas the clitoris of a great ape approximates the size of a fingernail, with variation, that of a modern woman approximates more that of a large apple seed, again with appreciable individual variation. With the vascular congestion commencing with arousal, this organ also swells. But why should it decrease in size?

The emergent problem that the clitoris faced in human evolution was that, whereas it had long evolved in functional design to terminate male entry as promptly as possible upon deep insertion, now a delay of climactic response became increasingly important. To remove the deeply ingrained role of prompt vaginal response would not be readily accomplished. The evolutional solution worked out for this problem was not to attempt to remove the existing organization but to overlay it with newly created processes. A hood elaborated, covering the clitoris, so fashioned that, while a full climax was held at bay, more gentle movements of the clitoris produced by surrounding tissues could produce mild stimulation and participate in prolonging arousal. In this and in other ways, the clitoris came to participate in new measures of delay concurrent with the neural inhibition being produced by frontal lobe elaboration.

Evolution also proceeded to support copulative delay by altering the functional anatomy of the clitoris. Upon full arousal, the hood and its nearby tissues become so enlarged that, in effect, the clitoris becomes anatomically retracted, rendering it, though still as sensitive as ever, more inaccessible than accessible. Furthermore, the location of the clitoris on the vulva moved upward, approximately two inches from the vaginal opening. In effect, these modifications mean that the clitoris does not come into significant engagement until late copulative action involving deep penile insertion. In spite of the fact that a full intensity of clitoral response is maintained, delay of heavy clitoral involvement thus becomes anatomically mandated.

It is clear is that during the human career the human clitoris, once so simple in its role, has become complexly programmed. Evolving human nature was profoundly needful of sufficient delay before engaging the reflexive nature of orgasmic responses. Before reflexive inevitability arose, other things now needed to happen. Step by step, as human intercourse elaborated it appears to have become both physiologically more intense and to have proceeded to involve more extended time. What originated primarily as a strong reflex action, an anciently programmed intense spasm, became more and more the culmination of lengthened physiological summation, a climactic event.

Over a period of time early in the human career the increasing strength of the reproductive forces that were then becoming experienced advanced the sexual experience to the state of an ever recurring adult need. Whereas a great ape female experiences brief climaxes many times daily for five or perhaps seven days every month, a mature woman may experience her far more powerful orgasms three times a week, or, if more nubile or needy, even as often as several times a day. In the service of advancing the reproduction of the species, human evolution has advanced sexuality from presenting as a periodic need essentially to the status of a maintenance need.

Regular intercourse reaching orgasm at human levels plays a supportive role in the maintenance activities of the individual. A favorable result of intercourse which proceeds through orgasm is its influence over feelings and behavior for a number of days afterward. A primary result of shared orgasm appears normally to be one of a sense of increased personal integration. As if there existed deep within the human brain a center needing certain levels of pleasure to affirm that life is good, the immediate effect of orgasm is a deep feeling of satisfaction. Depending upon its depth and quality, this permeates the personality for one or several days. This calming effect is often noted in the performance of the next day's tasks, which become more gracefully and easily managed. Typically the mental state is more composed. Basic security seems greater. Patience is improved, and even bodily coordination may become more graceful. It is an unfortunate commentary upon our day that so many women aware of orgasm's beneficial effects need to resort to vibrators to find their everyday activities less anxious, more becalmed, and more economically managed.

If two million years of human evolution were producing in women both a steady elaboration of sexual need and increasing levels of inhibition, what may have been taking place in men during this time?

The evolutional record supports that parallel changes were producing complementary behavioral patterns. The two main features of the human male's sexual advance appear to have been his experience of steadily increasing levels of sexual need and his increasing capacities for control. In the evolving human male increasing control was managed by frontal lobe advances, while in the female increasing inhibition was also managed by evolving frontal lobe developments. In a general perspective, however, there was a significant difference. Whereas female advances took place by new, fundamental transformation of function, male advances took place by steady, straightforward elaboration.

For fertilization to become accomplished, it is important that one sex characteristically assume the initiative. Having placed this role in male hands at an early hominid level, the subsequent course of human evolution appears to have continued to advance it, particularly through repeated elevations of androgenic action. It is perhaps somewhat difficult for civilized humans in today's more sedentary world to appreciate that the levels of physical activity that prevailed in plains life as ancestors evolved would have tended to have increased levels of sexual need, as well as to have repeatedly improved human powers of control. One can only note that one of the main features of the

American Indian observed with respect by early European settlers as they met the hunter-gatherer way of life on this continent was the red man's remarkable powers of self-control. The evolutional advance of the human male was consistently one of increasing sexual need leading him typically to assume the role of sexual initiator, accompanied by appropriate powers of control.

The fate of the experience of orgasm in man and woman appears to have been far from equal as human evolution progressed. In particular, what was originally an almost universal common experience of mutual climax came to realize numerous disparities. Since mutual climax was the point of origin for copulation at all earlier evolutional levels, this would have tended to prevail at the earliest human levels. The aim for mutual climax, when appreciated, still tends to prevail as an ideal and is often the accustomed goal. However, as voluntary behavior and personality features enter more and more into the extended human sexual program, simultaneous orgasm appears to become less and less a regular outcome. This variation of human response, noted particularly by cultural anthropologists, appears to have commenced with the elaboration of human cultures which took place after the introduction of agriculture, and has become ever more pronounced.

On the one hand, new interpersonal sensibilities have been emerging, rendering the sexual experience more meaningful. When intercourse has become sufficiently lengthened in emotionally aroused persons, it is not at all an unusual experience for the excitement of the rise to inevitable climax in one to precipitate the rise to climax in the other. On the other hand, in contrast to this delicate emotional coordination, there is the abundant data of modern medicine and cultural anthropology which report that the sexual experience of woman varies so enormously from one culture to another that in some women orgasm does not occur at all. Female orgasm is often simply unknown. It is difficult for one in today's enlightened climate to appreciate that in our own culture climax was not re-garded to be normal or ideal until the widely received researches of the latter half of the Twentieth Century. It was reported to be so rare in the Victorian era that physicians regarded woman to be normally anorgasmic. Even to many of the sexual cognizanti of the mid-Twentieth Century it came as a surprise that a woman's coun-tenance during orgasm resembled more acute physical agony than pleasure. That an enormous cultural advancement has continued to take place during the intervening decades is indicated by the fact that most men in American culture today prefer to prolong inter-course until their partner climaxes, while the majority of women have learned to climax at least part of the time.

In a broad comparative perspective, human reproductive evolu-tion reveals that orgasms have come to have multiple meanings. These are not identical in man and woman. For a man, orgasm is the incomparable mental experience that culminates his driven-ness to release his sperm, under favorable conditions transferring male gametes to female. Following this culmination of waxing ex-citement will follow a pleasurable sense of relaxation unmatched by any other experience of daily life, ideal as a prelude to sleep. Toward the woman who has facilitated this, he feels appreciation which, if deep enough, serves to bond.

It is often stated by culture critics that the age of exploration which commenced five hundred years ago has so progressed that today the exploration of human sexuality remains a last frontier of discov-ery. Historically, the nature and role of the human orgasm, particu-larly in woman, commenced to become physiologically clarified only in the latter half of the twentieth century.

For woman, the main satisfactions of intercourse, with orgasms or without, are typically emotional. The pleasures which arise with the intense exercise of her vaginal sensibilities speak yet for a typically deeper reward from male entry, a potential promise of care and loving affection beyond the sexual experience itself. Depending upon many variables – interpersonal compatibility, health, loneliness, fatigue, age, happiness, anxiety, level of sexual learning, and others – she may or may not climax. Although at times her sexual needs may dominate, more commonly, whether or not she climaxes, emotional gratification proves to be that most meaningful to her. However, particularly in sexually well experienced women, where such emotional needs are not significantly present, needs for sexual gratification may well emerge as a desire for their assist to daily living. Thus, what was originally a reproductive constant, female orgasm, has evolved in human evolution to a realm of total variability in its realization.

What, it may be wondered, do women regard to be most important in their relating sexually? In numerous surveys and study groups the routes to full sexual fulfillment in woman have received considerable scrutiny. Factors relating both to a woman's arousal and to her satisfaction of sexual need are now appreciated to run a wide gamut. These groups report consistently that neither the physical appearance of the partner, nor the level of the excitement that may be present, nor the mount of simulation from music, caressing, or even technique itself proves to be as essential an ingredient of the orgasmic experience as the quality of the emotional relationship that exists.

It would be an oversimplification to hold that androgens alone constituted the human endocrine advance. Undoubtedly, in male and female alike, clusters of hormones elaborated together, some of them estrogens. However, it would appear to be quite valid that the centerpiece of human endocrine elaboration proved to be testosterone. The diverse actions of this hormone match this emerging male role seamlessly. At the same time that testosterone is advancing the many physiological processes that enhance the male's physical capacities for meeting the larger environmental pursuits noted previously, it matures the penis, scrotum, and prostate during development and mediates the production of sperm at maturity. A very general statement would appear to be that, externally, testosterone advances male behavioral activity afield, while internally, at home base, it mediates sexual activity.

If an increase in male sexual activity existed as a potential route for incorporating the male into forming a closer family unit, more specifically, how might this have been accomplished?

A first observation would be that over the entire human career the role of testosterone appears progressively to have increased. This would mean that a motivational base for forming deeper sexual ties would have been steadily present. This increasing drive would appear to have been able to express itself in any of three elaborative ways. One would have been to increase the frequency of copulation. Another would have been to increase its duration, while a third would have been to increase its intensity. All three modes appear to have been employed in human evolutional sexual elaboration, although not quite all at once.

The simplest sexual advance, and probably the first to have been used, would have been an increase in sexual frequency with each episode remaining still brief.

A more promising route for improving male involvement, however, would appear to have existed in increasing the duration of copulation. The very pleasure of copulation would tend to prolong its duration. However, certain subtleties would initially present as problems. For one, female genital turgor may not have yet fully evolved evolutionally. For another, female climax would tend to occur rapidly upon male insertion. In this scenario, the first male sexual advance would seem to have been one of exercising measures of control during intromission and so acting as to avoid deep penetration. Up to this evolutionary point, male and female had experienced orgasm in synchrony. Now, with sexual skill and learning commencing to take place, mutuality of climax might not always occur. Significantly, however, with emotional development arising in the extension of a powerful vital need, in man and woman alike emotional processes were beginning to elaborate.

In time, additionally, intensity commenced to elaborate. As in physics it is axiomatic that nature abhors a vacuum, in biology it would appear similarly valid that, where properly poised, evolutional forces abhor missing a developmental opportunity. A curious and possibly counterintuitive new development appears to have occurred. Woman's new frontal lobe elaborations, in rapid evolutional development at this time, as previously noted, would appear to have introduced inhibitive processes that acted between male vaginal entry and the occurrence of climax.

The introduction of incremental inhibition meant that the female, in order to attain orgasm, would now need certain amounts of summative addition to previous levels of primed vaginal excitation in order to complete intercourse. The effect upon the male of such new female delay in attaining levels of arousal capable of producing an orgasmic response would have been to require him to perform increased amounts of intromissive work.

In the presence of increasing male sexual need this may have taken place automatically. As an added feature, increased vaginal activity would serve to increase the tissue turgor and thus to increase tactile sensibilities. The result would be an increase in intensity for male and female alike, the female from longer and heightened sensibilities and the male from longer and more potently driven forces required to overcome by summation the female's level of inhibition.

An interesting relationship appears to exist between the duration of intercourse and the intensity of the experience, in that longer duration tends to realize greater intensity. As long as the participants are healthy, longer durations tend to involve greater extents of summation, with this producing more intense orgasms. In this respect, the human brain appears to act much as an electrically accumulating energy system. Physiologically, while most energy accumulation at this time tends to take place in the synthesizing of tissues and in recurrent neural circuitry, in the human brain free energies appear to gather. The process of free energy accumulation that culminates in orgasm appears to begin with arousal, continuing with sensory summation for as long as the partners are actively engaged, gradually intensifying up to points of orgasmic inevitability. In its duration, energy accumulation deriving from genital stimulation may be supplemented by stimulation from such other body areas as the breasts, or at any point can be advanced by emotional forces. As points of high summative excitability are approached, additional stimulation from such regions may tip the scale to elicit orgasm, a total discharge of accumulated energies. The energies that culminate in orgasm are largely produced by increased striped muscular action which, upon orgasm, involves a relaxation of the total body musculature. In this programming, energies from sustained levels of excitement, still behaviorally amenable to some volition and desire, progress to points where all control becomes lost; the primal reflexive course takes command. Not all orgasmic events prove to be equal, however; rather, they differ according to the measures of energies that have accumulated. In general, the potency of the orgasmic experience proves to be proportional to the amounts of energies produced during the period of summation. To this, however, must be added the accumulations from emotional arousal which may have preceded vaginal sensory arousal.

In all of animate nature, the process of sensory summation which works to produce a human orgasm appears to be quite unique. Many examples of free energy accumulation exist in nature. These are perhaps most eminently noted in the electric eel where discharges are used for feeding and protection. A very common display of free energy accumulation and its discharge is observed in the bioluminescence of comb jellies by passengers in ships moving through ocean waters at night, or equally as common, in fireflies lighting a mid summer night. For reproduction, the phenomenon of human orgasm appears to be without match. Up to the level of the great apes, through inherent involvement of reward systems, animals appear to experience pleasure upon ejaculation, each appropriate to its level of evolutional development. None, however, appears to involve the grand accumulations of free energy that realize orgasms.

By guiding her sensations from copulative movements, shallow and brief or long and deep, as desired, the sexually learning and sexually learned woman becomes able to control her progression during arousal. As she becomes mature, through such variable pelvic action a sexually experienced woman becomes able to exert certain measures of voluntary control over the timing and the intensity of orgasms.

Comparative considerations of vaginal experience in higher vertebrates suggest a need for certain distinctions in description. The simple sensory summation of the reptile suggests a state described as an elementary climax. The uniquely intense experience of the human, which occurs after summation overcomes intervening processes of inhibition, suggests a more characteristic experience of orgasm. In such terms, the experience of orgasm, absent in the great apes, appears to be a distinctly human reproductive feature.

So appreciated, the phenomenon of orgasm appears to present a reproductive elaboration that developed during the human career.

Since delay between male entry and ejaculation became needed by woman in order to produce emotional bonding, the summative process, lengthened by female inhibition, came to produce more powerful contractile responses upon the ending of summation.

The progression of any given experience of intercourse is the result of a combined effort. The length of this interval usually extends from the time of male entry to ejaculation. Where simple impregnation is the goal, male duration can be brief, proving to be but a matter of seconds, as in evolutionally antecedent species. However, humans learn from experience that conjoint orgasmic responses are the most satisfying of all. Accordingly, higher male neural inhibition learns to exercise control. Such delay in ejaculation serves to produce emotional appreciation and shared personal identification. To the extent to which the male can exert control, the intensity of the female orgasm at the end of her summation will serve to elicit male ejaculation, in effect realizing the copulative synchrony most favorable for fertilization. In the process, human intercourse advances from realizing programmed reflexive pleasuring to deepening shared emotional gratification.

If the phenomenon of orgasm presents a distinctly human advance, at what evolutional stage might it have arisen?

Orgasms would appear to have begun to evolve early in human evolution, at about the time that frontal lobe developments through the introduction of inhibition were commencing to modify simple, more reflexive vaginal contractions. If not present at the very beginnings of the specifically human advance, orgasms would appear to have begun to develop not long after.

As the potencies of the human experience of orgasm increased, so also would human desires for privacy have increased. No sooner has a person experienced orgasm than a sense of privacy begins to surround its recollection; one has little desire to share it with others. The marked privacy that has come to envelop this experience today obscures the enormous importance that must have attended its steady elaboration over the ages.

XI

THE ELABORATION OF AFFECTIONAL SYSTEMS

Affectional systems involve activating dimensions of the personality which are pervasively present, subtly effective and much underappreciated.

*A*ffectional processes derive from emotional processes, with which they fuse according to levels of activation. Whereas emotions involve significant short-term mobilizations of energy, affectional systems are the subtle, long-lasting retentions of emotional learning which usually involve no special mobilization of energies. These processes exist in the realm of feeling. There they serve both to deepen the meaning of any given experience and as gentle directants from past learning orienting toward future action. Affectional feelings emerge out of the vast unconscious of an individual's memory into the immediacy of ongoing events. Marginally present in awareness but perfusing the mind, affectional processes are more apperceptive than perceptive. Sometimes described as fine emotions, these processes are numerous, in the description of William James as countless as the rocks upon a New England farm.

Neurophysiologists describe sensation, perception and apperception as components of the cognitive processes which give meaning to experience. Sensation involves the reception of energies by different modalities: vibration, sound, vision and heat. Smell and taste are chemical senses at the body surface. Perception involves the interrelating of various sensory modalities, producing an initial interpretation of immediate events. Apperception is the enlargement of perceptual meaning to include past experience and potential future courses of action. Of these three components, through the accumulations of experience, apperception in humans tends to involve the largest neural engagement. Reflecting this fact, physiologists often speak of the apperceptive mass, the large silent association areas of the brain which contribute personal meaning from the past to every newly experienced occasion.

Each of these cognitive components reflects an evolutionary tier in the composition of the central nervous system. Awareness of pain shows this stratification of meaning well. When a nail in a shoe produces pain, immediate thoughtless withdrawal is elicited. Such a reflexively well programmed organismic response to stimulation is basically reptilian. Interpretation and meaning are not yet incorporated, but tissue injury is present. Looking down to see the source of the pain, the person notes the nail in the shoe. The coordination of modalities in identifying the nail as the source is a mammalian type of action forming a perception. The shoe is picked up and looked at, an act which might be the response of an ape, who might then just throw it away. Reflecting, however, that the nail in the shoe was in an old pair too comfortable to be discarded, the human might decide to take the shoe to a cobbler for further repairs, hoping for future use in continued comfort. Such weighing of meaning is characteristically human, a voluntary act elicited by perception combined with apperception. In this, the larger portions of the brain's silent areas in the sensory lobes are contributing to provide individually meaningful interpretation to every event, in accord with that person's past experience.

Although affectional processes are certainly present in the great apes, they are much more highly developed in humans. This reflects their neocortical processing, the engagement of that portion of the brain which evolved the most during the past two million years. It comes as no surprise that the major source of human affectional systems is the human family, the primary context within which the distinctly human brain evolved. Such systems are produced by acts of warmth, kindness, and love, all of which abound in the home environment. Affectional processes are the enduring effects of loving experiences. Freud noted clinically that becoming loved tends to energize the entire personality. "By myself", Whitehead, the famous mathematician, once remarked, "I am a good teacher, but with Evelyn I become first rate."

Within the individual personality affectional processes tend to center mainly about those persons with whom there have been loving relationships. Mother, father, brother, sister and family members tend to be the most strongly formative, with mother usually foremost. The affectional role of the mother tends to persist most potently for the lifetime. It is curious how a deeply loving mother will remain a force within the personality of an adult son or daughter, encouraging that person on in life long after the mother herself has passed into the great unknown. Psychologists speak of the influence of such important persons as coming from images: a mother image, a father image or a spouse's image. The long experienced hunter-gatherer description of these lasting inner personages as spirits appears more fitting to the reality. They retain an enduring capacity to influence our thoughts and actions.

The apperceptive processes which are central for emotional memory fortunately benefit from general patterns which pertain for learning and forgetting. It is a feature of memory that the mind tends to forget incorrect items more readily than correct. In this respect, over time the mind can act automatically as a problem-solving and creative instrument. It is also true that, while remembering the strongly un-pleasant, the mind tends to forget the mildly unpleasant. Thus, while retaining recollection of that which is dangerous, by forgetting the un-pleasant, in effect it tends slowly to profile the pleasant. This life-for-warding feature mechanistic interpretation fails to explain. Memories of childhood, parents and friends become richer over time. When college alumni meet, incidents which at the time were initially un-pleasant or trivial are found to have become fondly treasured items for laughter.

Since all human learning courses through paleocortical emotional cen-ters before reaching neocortical levels for final processing, a summative profile of all of an individual's emotional learning presents an essential portrait of that person's working personality. However, within the com-prehensive emotional makeup of an individual the affectional processes involve only a portion, namely, that which relates to normal growth and reproduction. Hostility eclipses affectional processing, both that in memory and any which might be involved in immediately ongoing ac-tivity. Affectional processes buttress functional growth. In this perspec-tive, affectional processes are appreciated to be part of the cognitive composition and inner workmanship of the normally developing hu-man personality. A generous fund of affectional development serves to enhance personality growth, while a deficiency of affectional develop-ment tends to limit personal growth.

Deriving from emotional processes, which involve autonomic ner-vous function, affectional systems appear to be uniquely vertebrate, characteristically mammalian, and most elaborate in humans. Their evolutionary roots appear to originate in nest protection at the level of the fish. The brave stickleback attacks any intruder of any size. A small testy bird may often chase away one much larger which has tried to eat its nest eggs. Such actions involve evolutionarily early supplemen-tary energy systems for post-ovulatory reproductive function. At the level of the prosimians, the lemurs and the lorises, affectional systems are already well developed. Besides nursing each young, each lemur mother licks her pups and carefully combs their hair many times daily. For the latter activity, as one indication of the importance of grooming, it has evolved a special tooth comb on the lower anterior jaw. Great apes groom one another up to two hours daily, a behavior which results in the formation of bonds among various individuals and groups within the band. The affectional bonding which takes place in these various actions is mediated mainly by the physiologic involvement of oxytocin.

This pattern becomes much more elaborate in humans, where the increased frequency and duration of sexual activity itself usually proves sufficient to produce significant bonding. The major role of oxytocin, the bonding hormone, in primates involves maternal labor and delivery. At the human level the prolongation of intercourse serves also to involve oxytocin, eliciting bonding. Other factors being equal, once a pattern of regular sexual activity has become established, the attending forces of affection are usually maintained strongly enough to perpetuate the paired state. This appears to be a basic vertebrate feature as characteristic of birds as of humans. This action becomes somewhat qualified in humans, however, where sex has become so advanced as to need other dimensions to attain constancy, primarily more differentiated and longer acting emotions.

A crucial question for the formation of lasting pair bonding, often unrecognized, arises immediately after sex: What happens after sex? Here the role of affectional processes proves often to be determinative, particularly when a relationship is in its first stages. This specially holds for the fate of events the morning after. At this point the woman has usually become more attached emotionally; she has given of herself. Her feelings may say this, even when she may be attempting rationally to avoid it.

On the other hand, the man is usually in a refractory period. It is impressive how promptly after climax his mind tends to turn to other things. He is likely to turn to the problems facing him the next day. His mind wants to be free, not only from her but from all womanly entanglements. Basically, he needs to be free to leave home, classically to hunt for food. Freedom from immediate sexual need thus comes as a gift to him.

Whether or not he returns to resume sexual relations depends not so much upon the sexual experience itself as upon whether or not, and, if so, how much affectional feeling has developed.

For a woman to place personal demands upon a man at this time can often act to terminate a relationship which is tenuously developing. Difficult as it may seem to be, and however counterintuitive to her own nature, a woman's best response to her partner the morning after is to give him complete freedom to leave, with warmly appreciative undemanding memories. These he will carry with him. A man's sexually refractory period lasts anywhere from hours in a young man to days in one older. Once his distal tasks have become accomplished, his thoughts will turn toward returning. Coming to the fore, affectionate memories orient him strongly in his fatigued condition, producing desires to increase them further. In essense, the human male refractory period, forged by evolution, assures that he will leave home for distal activity, particularly hunting.

Though subtle, where the participants are emotionally open, the power of affectional systems for the formation of permanent sexual relationships is not to be underestimated. There are many women who happen to be not particularly endowed with sexually attractive features who, yet, by warm affectionate inner nature prove to be so appealing that their lack of sexual attractiveness in the long run does not matter. Whereas sexual attractiveness is surface, affectional feelings go deeper and last longer. It is a womanly talent, advisedly cultured, to enjoy affectional feelings in others, particularly those one finds interesting. Affectional feeling both enhances the experience during sexual intercourse and serves to perpetuate its meaning into the future.

Similarly, a highly affectionate nature in a partner can mean more to a man than certain levels of social disapproval. Many servicemen permanently based in locations overseas become sexually involved with local women. It has been not uncommon for many of these men to bring such foreign women home and to marry them. This occurs not only in appreciation of a given woman's sexual responsiveness but even more for her affectionate nature, which the man has responded to. Society has come quite to accept these different cultural couplings. Indeed, it could be maintained that affectional processes are essential for lasting sexual love. A woman who is too beautiful may fail sufficiently to develop affectional processes. She thereby may find herself handicapped in loving, an endemic problem of Hollywood. On the other hand, as a woman ages and her visual appeal lessens, well developed affectional systems serve to maintain a well bonded relationship.

Whereas the role of affectional processes in women tends to be fairly evident, their role in men is often quite obscure. However, male personality does not fully develop without them, and men with well-developed personalities often need the continued affirmation of affection to continue to perform at their best.

Perhaps no one becomes more aware of the powerful formative and activating role of affectional processes in well developed personalities than one of whom it might least be expected, the high-class call girl. Just as the lawyer sees human nature through sensibilities of fairness, and the doctor through appreciation of the healing powers of nature, the professional call girl, operating under the cloak of sexual privacy, sees human male nature in a bared, distinct light. No one senses more the drivenness of male sexual need when under deprivation, or the universality of such needs across all social, educational, ethnic, and economic strata, even well into senior years. Of the important role of affection in a man's professional nature, one discriminating call girl observed that, as strange as it may seem, the more power that a man has, the more he meets his goals and the higher he rises in his professional life, the more he needs to be embraced, to be pampered, to be loved, and to be kissed with affection. Many a high-powered call girl finds herself in the position of personal confidante for a lonely man. In such a situation the call girl has learned what many a man's wife has failed to learn - namely, that a lively womanly presence within a man's inner life has powers to enhance his entire personality in all that it does.

This is not to fail to note that the call girl's role, for whatever reasons it was entered into, proves to be emotionally telling. The absence of love in her serial sexual relations and her marginalized state in society incur deep personal costs which, continuing over time, produce personal deterioration. If she is able, she will remove herself from "the life" before it ruins her. Although she may see things other women may not see, her role is not one of either growth or health. Time is not on her side. Due to an absence of ongoing affectional reinforcements in her own life, she characteristically fails to take care of herself. Her own personality tends to deteriorate.

Affectional systems, a portion of the total apperceptive capacity of a personality, characteristically orient to the continued growth of the personality, relating eminently to processes of reproduction. As noted, it is the nature of fear and anger to eclipse affection. Hostile and fearful cognitive awareness tend to function dissociatively, with fear eclipsing sexual desire rapidly and totally. The presence of endorphins in sexual arousal tends to blunt both light pain and mild anxiety, but moderate anxiety or stronger pain also eclipse both desire and any immediate sexual capacity. Anxieties above certain mild levels tend to do the same. Every sexual experience with warmth or love, on the other hand, leaves its imprint in the affectional record.

These accumulate to form a mass of memories, more often silent than sensed. Though individual memories may fade, associated apperceptive accumulations continue, steadily enhancing the scope and depth of meanings. Who ever senses in a woman hugging her husband the father whose early tender love engendered hers? Yet, in a fine sense, he is there. Who senses in the accomplished business executive the mother who encouraged his learning and ambition? Aside from himself, who senses the elder, perhaps a teacher, who was the conscientious physician's mentor? Who can explain why the home team always has an advantage over a visiting team? In countless ways, deep within the person, affectional processes are ever at work, only occasionally recognized but always widely involved.

It is not uncommon for affectional processes, or their lack, to suggest explanations for the seemingly unexplained, such as may perhaps be noted in the morning news. The results of a study recently published in the Canadian Medical Association Journal may find explanation in such terms. This reported the percentages of men who arrived at a hospital emergency room for evaluation within six hours of onset of acute chest pain, a critical period for heart attacks. The result showed that 75% of married men met this criterion. In comparison, 71% of widowed men and 69% of divorced men arrived within this time period. Only 68% of the 4,403 patients in this study who remained single were so responsible. On average, married men got to the hospital 30 minutes sooner. They did so, furthermore, whether or not their wives were present with the occurrence of the pain.

Such reports are commonly amenable to varied interpretation. Without getting into issues of quality of life in marriage or divorce, a simple face value interpretation might be as follows. In the context of the present study, affectional processes subtly woven by a good wife into a husband's personality urge him to promptly seek proper medical attention under conditions of acute need. These exist as much in the immediate absence of wife as in her presence. If a man has become widowed, the powers of affectional proceses, though less, still prove to be significantly above the single state. Even divorced men may have residual affectional benefits in their makeup, though marginal. In a similar vein, a report in the daily paper that dog owners have fewer heart attacks comes as no surprise.

As personalities develop from childhood on, affectional systems form a bedrock of their nature. As these processes commence to form after birth, they are also the last to wane with age. Many never do. It is a common experience for the physician to note that as a person approaches the end of life all wishes and thoughts tend to fall away until they persist as one simple theme. A last wish at the end is to have close loved ones at his or her side. At this final hour affectional systems bridge the generations, providing an abiding sense of vital ongoing continuity. Through shared affectional feelings the human personality projects itself into the future.

Reflecting the fact that the brain is primarily an organ of growth, no particular nucleus or special pathway appears to be involved in affectional systems. If the anger system and the fear system are removed, growth-related affectional systems appear to have resources at hand that enlist the entire brain. Both the reward system and various activating systems appear to be involved, bathed in such agents as oxytocin, vasopressin, lactogenic hormone, and dopamine. These pathways appear to be part of our basic mammalian heritage, as evident in a well trained dog or a circus elephant as in a human. Affectional processes gradually enlarge, apportioning out their enhancive energies over the entire lifetime.

Varying enormously in levels of need and in levels of realization from person to person, from time to time and from one culture to another, the realm of affection, a higher realm of human personality development, appears to comprise a subtle dimension ever present in normal experience. Within the human psyche affectional processes work as silent gnomes, ever available to provide supplementary energies for negotiating ongoing developmental events. If human evolution should advance further, no more auspicious candidate for progression would appear to exist than any forwarding affectional systems. Such would further what appears to have been taking place eminently in the evolution of silent brain associational areas over the past two million years in contexts of familiation.

XII

HUMAN EVOLUTION'S UNCONSCIOUS FEMALE PLAN

> While on the one hand intensifying sexual arousal in early woman was producing ever more compelling desires for prompt consummation, evolution's unconscious workmanship was acting to prolong the duration of her summative interval.

*A*lthough polar ice formation had commenced several million years before, the major Ice Ages began 2 million years ago, introducing the Pleistocene Era. Reflecting newly recurrent variations in solar radiation, this era is estimated to have undergone over 120 distinct environmental cycles. Recurrent environmental instability, ever more severe, ranging from rainforest and lake to new desert, proved to be the context within which human evolution took place.

Such frequent and severe environmental changes prove to be highly consequential for existing forms of life. Among those that survive, evolution may take place with particular rapidity. While other hominid species failed, at least one ancestral hominin line coped during hardship and prospered during abundance.

Although increased levels of testosterone in males would return them home at day's end more sexually needful creatures, increased need alone could not produce the sexual constancy required for the more demanding care of the young then so important. Increased sexual need in the male could simply drive him to climaxing after shorter intromission. At the same time, finding herself and her progeny ever more dependent upon the male, an adult mother could find herself more powerfully driven emotionally, and thus likewise become prone to more prompt sexual consummation. As noted, physiological access to this was possible through more elaborate copulative experience. However, such an advance faced two behavioral problems. Increased male sexual need would tend to drive the male toward short periods of intromission, perhaps, as in great apes, as brief as possible. The female, on the other hand, would have been no less sexually driven. Finding herself now more dependent upon the male, she would also be well motivated to prompt and full consummation. How might evolution have overcome these behavioral problems?

The organ undergoing the most rapid evolution at this time was the human brain, and, within this organ, the areas developing most rapidly were prefrontal lobe processes.

Natural selection appears to have worked out solutions by genetic advances in prefrontal function. In the female, inhibitive processes were introduced between intromission and climax, slowing rates of excitation during accumulation, as if dampening somatosensory reception. Meanwhile, male frontal lobe processes also developed inhibitive powers capable of exercising control. As a result, the duration of intromission would have tended to become longer, with the climactic response able to become more intense. Such coupled advances, however, along with new variability introduced new complexities. If the man does not have full control, shared climaxes no longer result. In woman, the inhibitive processes when insufficiently overcome will block climax. When all works out well, however, such advances prove to be more satisfying for both. Such a program of development, it may be noted in the light of human evolution's rapid advance, holds promise of becoming readily self-forwarding.

If the aboriginal contract between the genders as sexual reproduction arose early in evolution involved the female's nurturance of ova and her attracting the male at times appropriate for mating with the male's production of smaller gametes and the responsibility for their delivery, such early human developments would have qualified as further advances of such a basic gender plan.

XIII

EXTERNAL WITHDRAWAL AND INNER WORLD EXPANSION

As sexual intercourse progresses, awareness of the outer world fades, while the inner world experience becomes ever more intense.

*M*ost of our daily life centers about external time, which coordinates individual experience with the comings and goings of others in society. As intercourse proceeds, however, the usually dominant sense of outer time steadily fades away. Less meticulously related to measurement, an inner sense of time relating to biological rhythms begins to dominate. First, security and comfort select a well removed location for outer world withdrawal. As the couple become more intensely focused upon each other in widening emotional and sensory expression, the outer world becomes ever more irrelevant. Step by step the external sensory modalities contribute less and less to immediate awareness. Eventually, awareness focuses entirely upon acute sensibilities of inner feeling. The sensory neocortical areas become eclipsed by prepotent limbic and thalamic mobilization. A unique state emerges in which, for a while, the world of ordinary experience seems entirely removed; outer time seems to be irrelevant as inner time reigns. A sense of timelessness prevails. After climax, as awareness of outer reality returns, a deepened sense of inner continuity may be felt.

If the larger external environment of hunter-gatherer society had come to be man's world, and the lesser realm about home-base woman's, the shared realm of the family at home base came to comprise a third human functional domain. The first intimations of the need for such a more withdrawn, private realm had commenced at early mammalian levels with the nursing of progeny by mothers. This often required their temporary spatial marginalization from the group. From its very beginning, nursing preferred quiet and separation from intense activity.

At the heart and center of the enlarging familial domain there further emerged over time a new fourth domain, centering upon shared sexual love and the interrelated emotional development of the parents. Mediated by finely developing neocortical processes, such a functional realm had never existed before. Indeed, pair formation in some other mammals and in birds often reaches levels which achieve reproductive pair constancy. However, the advanced complexity of the human personality, the developing interpersonal intimacy, and the presence of emergent culture characterize this as a uniquely elaborative human domain. Varying from couple to couple, from occasion to occasion and from culture to culture, this domain presents a realm still quite imperfectly evolved.

Although it is possible to debate sexual and gender inequalities in other areas of human life, to the extent to which this inner fourth domain becomes developed, there exists within it only a perfect equality of the sexes. With intromission, the sensibilities of the physical world and the cognitive identity of the self commence to fade. Here man and woman meet as one to one. Physically this union has often been described as "a beast with two backs". If a man has been working all day for a woman, with intromission at eventide woman returns the gift, and, in giving, receives her reward. The fact that orgasm in man and woman takes place alike at 0.8 second intervals per contraction suggests that for generation upon generation shared climax has been far more the rule for human ancestors than the exception. It would also appear that such an ideal response has enjoyed positive selection pressure. Human nature today yet retains an ideal synchrony of sexual response that civilized experience, in its haste and complexity, too often fails to realize, but toward which couples tend to orient. Although marked senses of privacy long shielded the fact from disclosure, anthropologists were able to learn in their studies of the Kalahari San that with intercourse the !Kung woman expected fully to climax regularly. Other studies of hunter-gatherer societies have indicated that, although this experience is far from universal, in this the San women were far from alone.

The result of a fulfilled inner life in woman – sexual, emotional and personal – proves to be a sense of deep, quiet appreciation of the man, while the mood in the man, similarly appreciative, becomes one of inner pride in being able to make his mate so happy. Very simply and of eminent practicality, this mode of mutual appreciation, immersed in sensibilities of shared equality, evokes desires in each to work for the other in an enlarging sense of wider selfhood.

XIV

MATE CONSTANCY AND
THE EMERGENCE OF THE
HUMAN FAMILY

Through improved bonding, sexual privacy made possible advances in intercourse that served to bring fathers
into the family constellation. Dual parenting made possible prolonged development of the human young and,
with this, more extensive environmental utilization.
Though moving divergently in their outer environmental
specializations, in home environments man and woman
came to experience ever closer interpersonal affiliation.

How ow could a sustained two-parent family have come about? The
fossil record does not offer a ready answer, or, indeed, promise any.
For the present, some answers may be sought in certain soft tissue
developments that have taken place between organisms at the level
of the great apes and the human. Human experience can also provide subjective information, the value of which for scientific purposes
must depend upon its objective validity. The prospective field for approaching this question is not ideal, but on the other hand, it appears
to be all that we have to work with.

For acquiring a first frame of reference, the distance between the position of a great ape mother and a modern woman may be compared. Although genetically the canopy-dwelling bonobo is slightly closer to humans, ecologically the edge-of-forest life of the common chimpanzee is probably closer to early human experience.

A great ape mother gives birth after an eight month period of gestation. She nurses her infant for approximately four years, during which time mother and child are in almost constant physical contact. Fearing aggressive males within the band, a mother and her dependent young tend to move about at its edges. This location, however, cannot be too far removed from the group, since the dangers of predators are always greater at the peripheries. While a mother is nursing one child, her offspring from an earlier pregnancy, not yet fully independent, is often present nearby. Adequate maturity for independent chimpanzee life begins to be realized at about seven years. Thus, a great ape mother may have one or perhaps two developing children under her care at any given time.

This simian reproductive program contrasts in many ways with that of a modern woman. Today's mother experiences a nine month period of gestation. She may nurse her infant for any amount of time, from none at all to as long as four years, or even longer when, as in periods of starvation, conditions for weaning are unfavorable. The number of developing offspring under her care at any time may range from one to a dozen, perhaps even more. Independent maturity is not reached until the close of the second decade of life. Requiring enormous parental work, the father is active in the family group, assisting in the care of children, although not always immediately present.

With the establishment of home bases, steady dual parental care became customary. This was predicated upon new levels of group safety and improved group comity. As noted in the chimpanzee, great ape mothers jealously hold onto their infant young, guarding them at all times. Human mothers, however, are willing to let other females handle their offspring. This gives the mother short periods of respite, while the child learns from early infancy to respond to many faces and to feel comfortable interacting with numerous individuals, some of them males.

Between these two evolutional positions, numerous other advances in parenting appear to have taken place. Whereas great ape females with infants often fear males, and thus keep their distances, in the new climate of group safety and of pairing privacy that emerged in humans such fears disappeared. New capacities for rational understanding, as well as new sensibilities of trust and affection in woman's nature, helped her to set aside her fears. Unlike male apes, who at times may kill infants, presumably to bring about estrus, human males do not practice infanticide. In this logic, perhaps the improved availability of sexual experience for early human males helped to bring about this safer female status.

In their reproductive advances, women broadened the scope of their affections to include attendant males. Warmly affiliative experiences, such as occur with grooming and intercourse, when often repeated serve to produce loving affection. Although more deeply interactive experiences may have commenced to take place at habiline levels, at erectine levels the new contexts of home base would have provided conditions for their marked advance.

When a woman gives birth, as noted, the same burst of oxytocin that induces contractions in her uterus tends also to produce a broader sense of altruism. She feels at once affectionate toward her newborn child and to the father who made the child a reality. Although the environmental dangers for an erectine woman were less than had pertained for a human ancestor at great ape levels, she would still have had reason to appreciate the protective role of the male. At the same time, she would continue to appreciate the male's provisioning of food, especially in dry season when it becomes scarce. With the birth of each child, the early erectine woman would have found herself in a state of increased dependency upon the male, in response to whose generosity her nature would tend to experience deepening appreciation.

The birth of a human infant tends to elicit bonding in the father as well, although usually less powerfully. A human child is born with features that tend to attract adults without regard to gender. A father tends to identify with a child, perhaps more if male, as an extension of himself. This process is both cognitive and emotional. In hunter-gatherer societies a newborn child is often seen as the spiritual reincarnation of a particular person recently deceased, a habit of thought not unknown to modern parents naming a new child. Whereas the mother bonds immediately, beginning with birth and nursing, the father often begins to bond more deeply later as the child develops, perhaps mostly through play. As the hunter-gatherer father turns homeward upon completing his day's chase, it is common for him to anticipate discovering what his child has learned to do during his absence. His attitude is one of both curiosity and expectant humor.

Similarly, experiencing inner emotional development in association with more extensive mating, a man comes to experience new feelings of tenderness toward a woman. As the sexual involvement of the pair increases, the nature of the sexual relationship changes from being more situational and irregular to one maintaining regular support of the family group.

Sexual learning and the forces of bonding often progress in different ways. Between persons basically compatible, love, though initially absent, may develop over time. It is not uncommon for a woman to enter into a sexual relationship with a man whom she likes but does not love, or for a man to enter into a sexual relationship with a woman whom he likes but does not love, only to discover that before long she, or he, has quite fallen in love with that person. With favorable emotional propensities and broadening experience, a woman in the throes of uninhibited intercourse, compulsively stroking her partner's face and hair, can find herself feeling as tenderly toward her mate as to her child.

As the evolutional record indicates, at early human stages males and females began to diverge in their environmental uses, as a result of which, before long, characteristic gender ecologies arose. In preparation for their complementary adult roles, gender differentiation commenced to take place during their developmental years. In addition to differences at birth in which boys tend to be more active and slightly larger, differences in the course of youthful development emerged. Human male and female growth commenced to diverge from the general mammalian course. What had been at all earlier evolutional stages a simple domed curvilinear progression of childhood growth smoothly tapering to level at maturity came to undergo modification. At puberty the introduction of a novel slowing of growth created in each sex a period of maturational delay.

The primary role of this developmental delay appears to have been the improved preparation of each human gender for its more demanding adult activities. In the male, lengthened maturation permitted improved mental and somatic abilities, preparing for wide ventures in the great outer realm. Upon the completion of this period of slowed maturation, a spurt of rapid growth at adolescence propels him soon to assume the demanding life of the hunter. At the same time, this extended period of home maturation improves his emotional preparation for later domestic life as a mature adult.

The slowing of maturation in a girl follows a similar course, but at more rapid rates. The course of her development for more proximal environmental and home roles appears to be less demanding physically and emotionally, as a result of which she becomes prepared for full familial responsibilities, on average, two years earlier. This physiologically well-ingrained difference in the number of years required for gender maturation still pertains for girls maturing today. Although actual ages at marriage differ widely, across all world cultures the age of the woman tends to average two years less than the age of the man. It is even more centrally a human characteristic that the age of the male at pair formation be greater than that of the female. For reasons sensed but unspoken and largely unconscious, man and woman tend to find it more comfortable for the man to be older than the woman, as well as taller. It is once more interesting to note that features of gender relationships formed so many eons ago should still be subtly woven into the dynamics of pair formation today.

In numerous ways, upon reaching maturity the functional ecologies of ancestral males and females tended to orient toward sexual constancy. Many features of the hunter-gatherer way of life direct toward this. When a man turns homeward from the hunt or from other work in the wider environs, he feels typically tired and often spent. Not by accident has the frame of the human male evolved in design for work. No occupation has been more demanding of male energies over the eons than the hunt. The capture of game can prove to be an easy venture, or it can be quite taxing, but the venture can rarely be counted simple. Travel distances can easily involve ten, twenty, thirty or more miles, while the hunting effort itself requires a demanding attentiveness. The about-face to home return involves a change in mental state. As he turns homeward, his thoughts promptly turn to rest, food, and sex.

The brain's inner reward system orients to sexual fulfillment. When the hunter afield anticipates a particular woman as his sexual partner, his neural reward centers become activated. Even such early activation begins to produce a certain initial sense of reward. Whitman well described this anticipatory state: "A woman awaits me", he said. "She contains all. Nothing is lacking."

To the extent that it may be present, often in the early stages of pair formation, romantic thoughts act as powerful forces orienting to sexual constancy. However, the well- committed relationship often proves to be capable of giving greater joy and pleasure than the stridently romantic ever can. Characteristically, both states want no third party involvement.

To the great benefit of the human family, sexual constancy tends to be self-perpetuating. A pair living in constancy will usually have met well the preconditions needed for satisfactory sexual coupling. These always present a certain hurdle against beginning to form a new relationship. Couples who have experienced sufficient sexual learning together note that, although the experience may become less intense, it gets better. Where the partners are compatible, excited years gradually merge into years that are delicious. In these, as male drivenness subsides, his control comes with greater ease, eventually becoming artfully total. At the same time, benefiting from her deeper sexual learning and a matured capacity for unreserved emotional openness, woman's responsiveness also becomes more profound. The result proves to be a lasting sense of shared intimacy. Clinicians often note the singular capacity of fulfilled sexual experience to reduce or to eliminate interpersonal hostility, a feature of human sexuality that well serves pair constancy.

Although human sexual nature, male and female, is fundamentally more constant than inconstant, it is not unalterably so. In the evolutional advance toward mate constancy, the positions of man and woman did not prove to be fully symmetrical. Basic differences were involved. Developing children benefit most from the attentions of both parents; the more complete the attentions, the better. Fully sensing the need for the best help possible for her progeny and emotionally desirous of no mate competition, a woman almost always wants a fully monogamous marital state. Evolving human nature appears to have reflected this by rendering any strong state of sexual bonding to be highly monogamous. Intensely bonded pairs, as noted, prove to be highly refractory to outside sexual attraction, as if hoarding something special, which, indeed, often it is.

The major problem with sexual constancy appears not to lie in the woman or even in the man but in the fact that both are members of a nomadic band, the numbers of which are always in certain flux. In humans at the hunter-gatherer stage, as in simians, the problem that arises for universal monogamous pairing is that, at any given time there tend to be larger numbers of adult females in the band than adult males. In such open and wide contexts, no woman can be left long unattended by male care. It requires to be appreciated that this problem has presented itself in human evolution for well over 800,000 generations, a period of time quite sufficient to have its accessory correlates become well ingrained into male and female biological natures. At every age, from birth onward, alike from susceptibilities to infectious agencies, accidents, poor nutrition, and other, males tend to die off at greater rates than females. Inevitably, in almost every band over the ages some women have become widowed at relatively early ages. Unless a better solution was found, many a younger brother has had to assume the responsibility for an older brother's widow and their children. Historically, for the satisfactory perpetuation of the group and its contribution to the future of the species, some adult males in a band, usually a small fraction, have had to care for two or more adult females and their progeny most of the time.

As a result of such common contingencies, the sexual inclinations of adult human males and females have evolved with significant differences. Though primarily monogamous, evolving human experience has molded the male mind to be prepared to take care of more than one woman. One result is that a man's psyche tends to experience an interest in the opposite sex on a much broader scale than that of a woman. Toward this, the male's heavy use of visual imagery and a nature generally less adept in fine emotions contribute. Although this male feature evolved to solve a recurrent band problem, in today's world it creates problems that neither culture nor further evolution has been able to solve. Half a century ago many a household included a maiden aunt. Today the problem of couple disparity arises in the context of the single parent. Since males and females age differently with females enjoying greater longevity, this problem could increase in severity in years to come.

Having coursed, similarly, over so many evolutionary generations, on a far subtler scale human female nature appears also to reflect the unpredictability of woman's reproductive fortunes in hunter-gatherer society. Should a woman lose a spouse for one reason or other, she will usually find herself in need of a new mate, at times urgently. Though primarily monogamous, a woman's psyche tends to prepare her for such an eventuality.

Though often only subliminally, her mind commonly harbors various affectionate and amorous inclinations toward particular others in the band. When conscious but mated, the level of such awareness in a woman tends to reflect inversely the strength of her emotional bonding. Strong monogamous feelings tend to obscure it, if not to totally obliterate it, while weak bonding permits these inclinations to find expression, even at times to flourish.

There is appreciable evidence of such widespread but marginal womanly inclinations. It is common for simple banter and humor relating to such subliminal tendencies to lighten everyday events. At deeper levels, these may become particularly noted when they emerge into awareness unexpectedly. It is the unusual marriage in our complex society today that does not involve certain measures of personal compromise. Over time, unfulfilled aspects of the personality, though sensed only marginally, may crave to be met. Few lines in English literature are more endearing than those of the staid, understanding husband in Noel Coward's play, *Brief Encounter*. A seemingly well-matched wife experiences a sudden, unexpected surge of romantic desire in response to a physician who takes a painful cinder out of her eye at a railway station. After they happen unexpectedly to meet each other subsequently on several occasions, the interpersonal encounter that seemed at first casual burst unexpectedly toward an episode of adultery, prevented only by a stroke of fate. Upon sensing his wife's quite uncharacteristic inner turmoil upon her unusually late return home, her husband gently and somehow knowingly says to her, "Thank you for coming back to me."

Human evolution and subsequent cultural development have worked out partial solutions to these reproductive disparities. Since males die off at greater rates than females at all ages, 106 males are normally born for every 100 females. This reflects a biologically well-forged trajectory, designed to realize at early maturity the ideal state for adult parental function, one female for every male. Totally beyond any awareness, every newly formed couple faces their future with a parental state optimal for the nurturance of prospective children.

On a far less exact but no less exacting scale reflecting the enormous evolutional history of the emergence of pair constancy, human nature tends to end promptly sexual relationships void of emotional or interpersonal compatibility. This strong bent also serves the interests of human young yet unborn. The development of strong dislike toward a particular sexual partner, particularly by a woman, tends to eclipse sexual desire. If this extends to her sexual refusal, the man, especially, will want strongly to terminate the relationship. When this happens, hopefully it occurs before children have been produced.

Such a scenario, however, involving strong emotional processes, can take place at any marital stage. As the hot season begins among the Kalahari !Kung, the band breaks up into family units. Each family must fend for itself in the oncoming dry season when food becomes scarce and tempers are sorely tried. When the rains return, family groups gather again. This is a time of jubilation, in which unhappy spouses, particularly those recently wed through parental arrangement, may leave their coupled state and return to their natal households. Similarly, when a wife is distressed from overwork, it is not uncommon for a woman in hunter-gatherer societies to request her spouse to take on a second wife to help, ideally, when available, a younger sister already compatible.

Any sensory pleasure enjoys an initial period of freshness, following which, in spite of constancy of stimulation, the intensity of the experience wanes. Sex is no exception. Even in the most favorably matched couples the sexual freshness of early experience gradually wanes over time. This may become apparent after two good years, or perhaps much sooner, or perhaps later. Especially in today's world with its difficulty in maintaining lasting friendships, it is essential for the forming of an enduring relationship that the dimensions of emotional understanding and interpersonal appreciation become sufficiently developed. As the sense of special sexual pleasuring, becoming more commonplace, loses its glow, the extent of these inherently binding dimensions will most likely determine whether a given relationship lasts or fails. The wise couple freshly enjoying the intense joys of sexual harmony together works from the beginning to enlarge the features of their shared life and personalities that work so effectively to realize permanence and lasting sensibilities of goodness.

Since becoming evolutionarily consolidated, the nuclear family has remained the basic unit of all subsequent higher forms of human organization, including those in all modern cultures. If the rise of sexual privacy presaged the emergence of the human family, the long continuation and expansion of sexual privacy over the years has led to endearing enrichments within family culture and, with this, to the deepening of human personality.

XV

HUMAN ADVANCES
IN SEXUAL INTERCOURSE

Human sexual arousal involves a comprehensive syn-
drome of sustained metabolic effort that is unique to
the human species. The human sexual experience
involves advances in anticipation, in the intensity of
arousal, in its duration, and in its after effects.

A major tenet of this study holds that human evolution advanced
reproduction to a new order of functional magnitude. A corollary to
this is that the biological work performed in realizing the species'
reproduction markedly increased, both in the proportioning of the
systemic work involved at any given time and in its cumulative lifetime measure.

The evolutional extent of the human reproductive advance is not immediately apparent. Anatomists comparing the structure of the bones
or of other organs of a human with those of a great ape note such
features as longer legs, shorter arms, and a larger brain case, all usually deemed insufficient, in sum, to qualify as a new cladistic family. If, however, the claim is made that this species has so "adapted"
as to have mastered the environment of the planet, and has so advanced reproduction that, arguably, it has become heavily involved

in reproducing all of its lifetime, a claim to special status becomes more substantial.

In the course of human evolution, as frequently noted, a crucial land-mark appears to have been passed when ancestors changed from being victims of predation to being safe, themselves having become preda-tors when needed. At this point, access to prolonged copulation would appear to have emerged, opening for the first time a veritable new uni-verse of inner experience unknown to any other creature on the planet. In well advanced state, this realm appears still to be evolving today

Human advances in reproduction were comprehensive. These would come to include, in time, the lengthening and intensification of inter-course, a more profound experience of orgasm, sexual learning, the deepening of emotion, the personalization of the act, and discrimi-native mate selection. While such peripheral organs as the penis and vulva were evolving modestly, the central nervous system was evolving dramatically. Although moist of the central nervous advance related to external environmental activity, significant internal elaboration was also taking place. During this period human personality developed significantly as familial emotions broadened, many through subtle refinements of feeling. While intercourse became more intense and more emotionally involved, it was also coming to take place ever more frequently.

It was also suggested earlier that in ancestors entering woodland life a selective lengthening of body hairs accompanied by a more gen-eral hair abiotrophy may have evolved in hominid parents, aiding the transport of infants and children in dangerous locations. If such long hairs were still present at the beginning of specifically human evolu-tion two million years ago, with the establishment of home bases they would have been no longer needed.

It is not unusual for features undergoing evolutionary change to find more than one new use. Aside from the fact that long body hair was no longer needed for child transport, and thus would tend either to atrophy from disuse or perhaps advance to become again the heavy protective coat of other plains animals, certain other opportune functions of total hair loss would have been making their appearance at this evolutionary juncture. In particular, along with needs for tracking and hunting skills, new thermoregulatory needs and new major advances in reproductive function were emerging.

A new male role made possible by total hair loss emerged, aiding hunting and tracking. Although full confirmation has been elusive, for many years it has been widely supported that the loss of truncal hair in humans occurred as an adaptation to the hunting way of life. The simple common sense hypothesis holds that increased capacities for sweating over large body areas improved human exertional abilities, particularly enhancing close action in a chase.

Reservations concerning this simple common sense hypothesis have been several. All mammals on the plains have heavy coats of hair that protect them from direct sunlight. A bare body exposes it to increased heat loads, requiring greater amounts of sweating for inner thermal stability. How could humans ever have managed to add new exertional expenditures to the already heavy costs of existing thermal loads under the direct sun? Sweating, it is also noted, does not usually take place in humans until well after the critical early minutes of a dash. The facts that man is at once the most water dependent of all large animals on the savannah and the species with the least water storing capacity compound the perplexity.

In the light of earlier considerations in this study relating to human hair abiotrophy, an important consideration in addressing this problem would appear to be that humans entered their early transitional environmental period some two million years ago in largely hairless states. Under such circumstances, early humans had no choice but to begin to scavenge and to hunt for food as they were.

It is often said that evolution's method is simply anything that works. Human sweating is noted to have developed a range of capacities. For light amounts of work sweating occurs at rates producing up to 1 liter per hour, for more sustained effort at rates of 2 liters per hour, and for heavy effort over brief periods at 3 liters per hour. These rates are singularly human; no other creature matches them. Humans gained these capacities not by increasing the numbers of their sweat glands but by increasing the production per gland. What evolutional necessity worked out for early humans was not so much a special capacity to expend heat through high evaporation in a spurt of extreme effort as a special capacity for long term moderate work under circumstances of increased external heat. What is generally described as hunting involves work that is expended primarily in long range efforts of scouting and tracking. Given the environmental context, this appears essentially to have been what early humans primarily needed.

It is difficult for moderns to appreciate that, though he is more likely to use bow and arrows, a Bushman can capture an impala through tracking alone, using no weapon more than a well pointed stick. Should the hunter discover the fresh tracks of a kudu or other of many antelope species on a morning, he knows intuitively that if the animal can be found before the midday heat passes he can probably track it down successfully. Under the heat of midday, furred animals seek shelter in shaded areas. Upon sensing an approaching human should a sheltering animal dart away, the hunter still knows that with further pursuit the animal will wear down before he himself will. Eventually he will find an exhausted, heavily panting animal, too fatigued to run away, still likely to try to fight but essentially helpless against even a club. The critical feature in the human hunt is not speed but endurance. In spite of having the least capacity for the intake or the storing of water among the large mammals on the plains, humans have evolved to become the species with unmatched physical endurance.

A second outcome of complete body hair loss at this evolutional juncture would appear to have been the facilitation of the change from group grooming to the practice of grooming by coupling pairs. Although the human skin at this time may have been becoming more sensitive under new endocrine advances, hair abiotropy would have made it even more so. An abiotropic hair follicle might not be any more sensitive than its larger predecessor but the absence of a pelt of long hairs means that an ordinary act of touching came to involve both hair and skin surface alike. The mental cognizance of a gentle hand upon a recipient's skin thus becomes more stimulating. The new breadth of total body exposure and a likely increase in skin oils render touching, ever the most potent of the sensory modalities, to become all the more effective in the forming of deep personal relationships.

Several interrelating functional advances were emerging in the human scene at about this time. Touching the skin serves to release oxytocin, the bonding hormone. Evolutionally this had commenced to elaborate first in mammalian mothers as they began to nurse their young. Occurring in both giver and receiver, the amounts of oxytocin released depend upon the extent of the grooming and its duration. Grooming in great ape dynamics is a major cohesive force. Oxytocin produces warm affiliative feelings, with inclinations toward identification and trust. When sufficiently developed, bonds of affection so formed realize varying measures of permanence. That the extent of such grooming was more than trivial is suggested by the fact that great ape grooming normally consumes up two hours daily.

At about this time, presumably early in the habiline career, the feature of sexual inhibition, so fateful for the human future, was beginning to develop in women. Such inhibition could not have evolved had not ways to overcome it developed concurrently. The increased role of personal grooming at this evolutional stage, serving to facilitate sexual arousal, would have provided a needed source of complementary stimulation. Thus, through more intense and personally focused grooming over large body areas, hair loss would have been able to facilitate the emergence of human pair bonding.

The expanding role of the sense of touch in the widening human sexual agenda now involved, first, variously eliciting, and subsequently of incorporating affectionate feelings relating to copulative experience. It is appreciated that in human sexual arousal almost universally certain levels of production of oxytocin occur. In these rapidly developing tactile functions, stroking, caressing, fondling, gentling, and kissing present tactile elaborations that elicit emotional feeling. Experienced as affection, when they become stronger such sensibilities progress to feelings of love. As early humans entered their new evolutional scene, affection began to emerge as the normal context of feeling within which sexual intercourse was coming to be experienced.

A major, seemingly unrelated behavioral change that appears likely to have taken place early in human evolution would have been a preference for frontal entry over posterior during intercourse. Where attention to the surroundings previously had been necessary in order to keep watch for possible predators, posterior entry had been essential. However, with fears of predators removed, frontal entry became newly possibile.

Why, however, would such a change take place? It is reported that posterior entry is more efficient for achieving fertilization. For many experienced couples today posterior entry is reported to be the clear position of preference. A change from posterior positioning to frontal would seem not particularly to have been needed.

The critical feature in regard to making such a change would appear to have been that frontal positions present greater sources of accessory stimulation for female sexual arousal. In order to attain a mutual climax as intercourse became more prolonged, coupling pairs had to work more diligently in sensory summation to overcome female inhibitive delay. In posterior position, a woman's arms are of little use, and a man's use of his arms proves, similarly, to be quite limited. All facial communication is compromised. On the other hand, in paired frontal position all hands are completely free for shared exploration , and all facial features become fully accessible.

It is possible that certain evolutionally well-established aspects of dorsal and ventral body function in higher vertebrates may have been relevant to such a change. Whereas the mammalian back tends to be presented to the external environment in protection against forces that might produce injury, mammalian frontal areas appear to be used more often for reproduction. Tactile sensibilities used for infant care are greater ventrally, and mammalian nipple lines are universally positioned frontally.

The anatomies of human male and female genitals today suggest that, for whatever the reason, a fundamental change to frontal position during intercourse has been present for a very long time. In its erect state the penis has advanced from being straight to having a mild upward curvature. At the same time, in shared accommodation, the vagina has modified its position of entry from being located slightly posterior to become clearly anterior. The anterior location of the G-spot suggests that frontal entry was not only common but preferred. Variations in paired copulative positioning may be mainly of more recent emergence, particularly relating to the enormous cultural diversification that took place after the discovery of agriculture.

It is difficult to know which among the various anterior sensory modalities would have been most helpful for producing arousal. Perhaps the question is spurious. Very likely, each person's exchange was individual, involving the use of any one, several, or all, in different proportions by different persons at different times. With increased duration, sexual learning commenced to occur, through which, over time, copulation became individually creative and personal. It seems more fitting to individual experience that patterns of arousal developed in accord with the learned ways of coupling pairs using various modalities. With varied frontal stimulation human intercourse was commencing to become creative.

Eye contact has evolved in modern humans to act as a silent language, used more subliminally than consciously Sustained eye contact can prove to be deeply moving. Among other things, human eyes are used to communicate, one to another, a state of inner sexual need. Should a woman desire, she can use her eyes to seduce a man - as could the classic Aspasia in ancient Athens at a distance of twenty paces, as did the inn maiden to Tom Jones in the first novel of the English language, and as today many a visitor to a bar will readily testify. As intercourse commences and desire intensifies, warmly appreciative eye contact generally pertains until, as climax approaches, the eyes tend often to close.

In states of sexual need, the human face may come to speak a language of its own. Desire tends to produce a warm, intensely focused attentiveness. As the experience of intercourse progresses, a woman's facial features tell the man how aroused she is sexually and emotionally, whether she feels warmly close or has become inwardly consumed, particularly as climax approaches. At climax, with contorted face, she is often deeply in another world.

In frontal position, simple words can have more potent effect, and soft words become more readily exchanged. As the mutual program progresses, words from the man or woman may give way entirely to mere sounds – moans, brief exclamations, and even at times yells. These crude sounds expressive of feeling prove to be more deeply telling than words. Having arisen in evolution long before words, many such sounds are characteristic of the sexual act itself. A third party hearing such sounds at a distance, though receiving them far from clearly, knows instantly their meaning.

It is reported that in some human cultures kissing is absent. It is difficult to believe, however, that kissing has not been an integral part of the frontal sexual exchange for most of the human career. In chimpanzees, light lip touching is often noted, while the human sensory cortex gives enormous representation to the lips and face. In our own culture, kissing during arousal tends to be quite taken for granted. In a fresh experience it is often intense and continous. In shared appreciation upon a deeply experienced climax, warmly appreciative kisses are almost universally exchanged.

In full human arousal, a sex flush commonly takes place in woman, giving her a certain radiance. This covers her face, her neck, and her frontal chest. These areas become more sensitive to touch. Included in this vascular surge, the breasts become slightly swollen and more sensitive. With excitement, the nipples typically erect. Many women desire breast stimulation as part of their arousal or to aid in their sensory summation. Some claim to be able to experience orgasms from such stimulation alone. Many men assume that a woman's breasts are highly sensitive all the time, but this is an exaggeration. Most of their particular sensitivity occurs with arousal, for which these tissues are usually in a primed state.

A silent human sexual advance during arousal takes place through an increase in blood pressure. Compared with anything subhuman, the human state of arousal involves an increased blood flow to the entire skin, the brain, and the genitals. The entire body surface becomes more sensitive, tending to sweat more readily. The mind proceeds to become inwardly more intense. Elevation of blood pressure and increased depth of breathing sustain the elevated metabolic state as it proceeds through summation to accumulate energies for orgasmic climax.

Besides modest external genital organ development, the advance to human levels of reproduction witnessed the development of special hairs in areas of body friction. Reflecting a new, high order of use, these pads protect areas of high pressure contact from abrasion. As a woman's arousal deepens, her pelvis often participates with rhythmic thrusting. Whereas male pelvic movements tend to be more linear and longer, female pelvic movements, using greater hip mobility, tend to be more circular, shorter, and more varied. Thus the patterns of male pubic hair tend to be more diamond shaped, extending upward on the abdomen toward the umbilicus. Those of the female tend to be triangular and are localized in the pelvic region. With more focused pelvic movements accompanied by sexual learning, a woman becomes able to attain more effective clitoral stimulation, and becomes thereby able to guide her progress toward orgasm.

Among the diverse sensory pathways that have developed in human evolution that are capable of enhancing sexual arousal, foremost, as noted, has been the role of the clitoris. As human nature has evolved over the past two million years, an initially rapid, gross clitoral reactivity progressed to become generally milder with a reduced reactivity, making it capable of prolonged excitation. Upon sufficient satiation, a gross, evolutionarily typical, highly reactive phase secondarily proceeds, producing a prompt orgasm. This often marked lengthening of response was accomplished anatomically by modification of attached clitoral structures. These formed an overlying hood that attached to the labia, all of which have evolved during the human tenure to levels of subtle sensibility that contribute to sexual arousal.

Whereas human sexual and gender differences in Western culture have undergone reduction in emphasis in recent decades, in numerous traditional cultures over the years these differences have seen ornamentation. A physician traveling deep into Africa today, particularly in rural areas, would probably still be able to find in some women labia minora that have been intentionally enlarged. As preparation for marriage at puberty, these young women have been taught by older cousins to manually stretch the rapidly developing labia minora by performing downward traction between the thumb and forefinger daily each morning. In fairly short time, each labium may develop appreciably. Although an unstretched labium may at times show a crescent shape two to three centimeters in width, stretching may widen each lip from three to six centimeters, and occasionally in some conscientious young women to more. In congested state these enlarged labia produce a cornucopia effect with the vagina. This modification, particularly appreciated by men in sexual decline, serves to reinforce sexual bonding. If a woman wishes to prevent male entry, she can do so by tucking these ordinarily flabby loose tissues into the vagina. Once puberty has passed, this modification becomes so difficult to produce that it is no longer attempted.

Not only do the breasts and perineum become more sensitive during arousal but also frontal body surfaces. A traveler deep into Africa a century ago would often have noted in some tribes women with nodular scarified markings on their frontal frontal chests and abdomens. These were produced by teaching them at adolescence to prick their skin with needles dipped in ashes from the fire, a practice that has disappeared due to its painful nature and to the subtle effects of cultural change. The purpose of such nodules, typically arranged in broad linear patterns over frontal areas, was to enhance tactile sensation during intercourse.

At the same time, in a practice that has probably not disappeared, the young woman was taught at puberty to stretch her nipples. This, she was advised, was to assure that infants at birth would have no difficulty nursing. However, at the same time, in males and females alike, nipple erection is appreciated for its role in sexual arousal.

A general result of sexual learning in the human reproductive program is that the sexual experience becomes personal. This feature, mediated by the neocortex that was concurrently evolving, has multiple aspects. As many women describe it, there is a certain "before and after" to vaginal experience. If a man has once entered her, the barriers for a second entrance have become much reduced. With extension of intercourse, an unspoken sense of an intimate knowing, a shared nonverbal understanding tends to develop. This tends to occur with or without climax. Learning how to please each other sexually becomes a personal focus, developing in each a characteristic style.

Irregularity in the experience of climax appears to have been a major consequence of the extensive reproductive elaboration that took place in humans, with its inclusion of sexual learning as an aspect of maturing. Whereas in the male climax is almost universal, as a result of her incorporation of inhibition and delay orgasmic climax in woman has come to be far from universally experienced. When it occurs, the magnitude of her experience depends upon numerous variables. As previously noted, these include her level of immediate need, the duration of sensory summation, the level of her emotional arousal, and even such background features as health, age, or fatigue.

Indeed, the main feature of the incomparably advanced state of the human reproductive program, it may be argued, is the wide variety of responses which may take place in woman. Like other basic needs, sexual needs may be deprived, in which case they commonly intensify, or they can become sated, in which case they may diminish markedly for a while. On any given occasion, a woman may climax or may not climax. Should she climax, her response can vary anywhere from mild genital spasms to total body convulsions. One might expect that evolution would have perfected a complete, universally fulfilling womanly response. This remains true in an ideal sense. However, it must be recorded a unique biological feature that, unlike all other female vertebrates who tend to experience a shared climax, one aspect of woman's unmatched reproductive complexity is that an experience of intercourse quite commonly takes place with no orgasm having occurred at all.

Certain questions arise. Why, it may be wondered, are there so many routes for human sexual arousal? Could not sexual arousal have been more effectively accomplished by advancing one or two routes to perfection? Why is woman's response so variable? Does not the sheer complexity of the female response contribute to its irregularity in attaining fulfillment?

It has been supported above that the reason for the engagement of so many pathways to arousal was the need to overcome inhibition that was developing in woman's reproductive nature. Indeed, on review, it would seem that every possible pathway available was enlisted. This suggests a certain evolutionary urgency that lasted for a significant period of time, consistent with the enormity of brain development that was then underway.

Woman's irregularity in response appears in all likelihood to have been an unavoidable consequence of her emerging reproductive complexity. There were, simply, many new ways that this new program could go wrong. One consequence of this novel complexity, as noted, was the requirement of increased male work. This would take place, it may be presumed, provided that such new demands occurred incrementally, never individually too large. With more pronounced orgasmic experiences, such consequences would usually tend to become appreciated by participating males.

In spite of such wide variability of response, there is appreciable evidence that a woman has greater potential sexual capacity than a man. In comparison with the male's strong, more overt and generally more direct erotic nature, woman's capacity for sexual response, though less immediately apparent, appears to be both deeper and more varied. Whereas a man may rarely experience more than one climax upon a given occasion of intercourse, many a well attuned woman may experience several, each often more potent than the one previous. This appears to relate to the proportionately greater physiological endowment in woman for the species' reproduction, and to the ecologic necessity for a refractory period in a man by which he may depart from home without feeling inner conflict.

Perhaps certain corollaries of the evolving human reproductive advance may offer clarification. The organ that was undergoing the foremost development during these years, doubling in its size, was the human brain. The preeminent role of the neocortex, evolving the most extensively, was the advancement of external environmental adeptness. However, it is widely appreciated that a common accompaniment of improved adaptation proves also to be improved reproduction. While the neocortex was enlarging rapidly, at far lesser rates the paleocortex and the thalamus were also enlarging. New preliminaries were emerging before satisfactory sexual intercourse could begin. Such negotiations range from selecting a partner, to choosing appropriate locations, to meeting the specifics of sexual arousal itself. Once arousal has commenced, stronger emotional forces and deepening sexual needs take command. Thus, while emerging humans were broadly developing fine emotions relating to the human family, in more aboriginal ways they were becoming, as well, more deeply sexual creatures.

Emotions cannot always be counted upon to bond permanently. As the newness of sexual and emotional experience wanes, evolution came to count upon the interpersonal involvement of the couple to insure more lasting coupling. In precivilized societies, where specialization and complexity are not yet as highly advanced, the main source of parental bonding during the third and fourth decades usually takes place in their shared attentions to children. Until modified by human cultural development and by civilization over the past ten thousand years, this working context pertained for an estimated 99% of the human evolutional career. In civilized contexts, personality development and mutual compatibility tend often to assume predominant roles.

The most articulate descriptions of sexual gratification, as those of any strong need, tend to occur when they are met after severe deprivation. A plain glass of water never tastes so refreshing as when one is thirsty on a hot day. Confronted by the impossibility of any adequate description of a first full orgasm, some women with loyal, loving, and patient husbands, have wanted, nevertheless, to try to convey to others the depth of their newly discovered experience of orgasm. After years of initial frigidity, the onset of a first total orgasm was described by one woman as like going over Niagara Falls in a barrel. It was an experience of such beauty, said another woman, that just thinking about it brought tears to her eyes. Another woman described it as a mounting symphony of intensifying rhythms, swelling uncontrollably to an inexpressibly grand crescendo.

Women who are more accustomed to realizing generous climaxes also describe their orgasmic experience with deep appreciation, though usually more prosaically. They are more likely to describe an intense pleasure waxing to rapture or ecstasy, or an experience of melting into a dream. A total release of the self leads to sensing a realm at once wondrous, timeless, and marvelously becalmed. For many a tense and tired woman in today's harried workforce, an orgasm at day's end restores an inner sense of well being.

In a deeply fulfilling human sexual experience, the ordinary sense of the individual may undergo transformation. As intromission lengthens, the sense of self expands to involve two persons. Upon its completion, the two are no longer quite the same. Each feels deeply appreciative of the other. For a while, a level of selfless objectivity may prevail, with a clarity of mind rarely, if ever, attained elsewhere in ordinary life.

With fulfilled loving, other personality changes often take place. It is an interesting clinical fact that needs for self-respect do not become fully met until needs for loving have been fulfilled. These changes may or may not be evident to others. The woman whom a man loves becomes more beautiful, most of all in the eyes of her lover. The man whom a woman believes to be wonderful becomes more wonderful. When afield he becomes a better hunter, and at home a better father. Although spoken words are effective, for deep effect they need often to be but few. Personality changes commonly take place without words, affirmed alike by gestures, knowing glances, or perhaps mere touches in passing.

In such a kaleidoscopically changing world as we now live, such highly rewarding sexual experiences cannot usually be sustained consistently over long periods of time. Couples tend to follow more humdrum routines in which sex becomes more a maintenance activity. Equally, however, such experiences are rarely forgotten. Among other things, the deep inner gratitude which these emotionally fulfilled women experienced, had the capacity, they discovered, to turn ordinary tasks of daily life into pleasures. It is as if the entire mind and personality somehow had become reset at higher levels. If every sexual experience largely reflects the energy that was put into it, each highly fulfilling experience emotionally redefines life by an increment. If any measure of the evolutional advance of human reproductive processes were desired, such attempts to describe the experience of a richly fulfilled orgasm and its unique aftermath would certainly qualify. At simian levels nothing faintly resembling such profound experience exists.

We are living today in a world of uncharted sexual exploration. Our recently acquired knowledge of sexual physiology is quite without historic precedent, as are our stunning medical achievements relating to human reproduction. By any quantitative measure, the sexual experience of younger generations today seems light years beyond that of most seniors today. But issues are many. Our problems in the sexual realm derive not from deficiencies in our basic sexual nature but from our failure to match advances in sexual understanding with advances in emotional understanding. We are failing to integrate dimensions of sexual experience with longer range interpersonal compatibility. In a world where personalities have become highly specialized and complex, enduring relations sufficient to maintain the family have become more and more difficult to realize, so much so that it is becoming a question whether the family itself will survive. Advances in emotional and interpersonal understanding have not kept pace.

Although we are advanced well beyond hunter-gatherer culture and it would be impossible to return to the ancient ways, perhaps a better understanding of where our sexual nature came from may help us to see our way forward into a happier future.

XVI

WILL ROMANCE DISAPPEAR?

Romance, art in gender relationships, has deep evolutional roots.

*I*n today's rapidly changing world, the question frequently arises whether, as often seems to be happening, human romance will disappear. Cultural trends seem to confirm the clinical studies both of Kinsey and associates and of Masters and Johnson which support that romance is not an essential part of human sexuality. Many cultural trends suggest that romance is disappearing. For a large portion of the millennial generation romance is not sensed to be in the cards. On the other hand, our cultural heritage is rich in expressions of romance, and has been so for centuries.

Although it is not clear at what point in their relationship Sir Philip Sydney's beautiful sentiments were expressed, one may wonder whether any since the sixteenth century describe the essence of romance more simply or deeply. As stated by the woman, his sonnet declares:

> My true love hath my heart, and I have his,
> By just exchange one for the other given.

THE EVOLUTION OF HUMAN SEXUAL PRIVACY

In essence, these plain lines describe a well-developed state of mutually shared love. Are such richly fulfilled sentiments disappearing?

There is a tendency to think of romance as a fragile invention of civilization. Romance seems particularly to have flowered with the coming
of the Renaissance. One immediately imagines courtly troubadours
singing to fair ladies below their balconies, or recalls the classic lines
of Ben Johnson:

> Drink to me only with thine eyes
> And I will pledge with mine;
> Or leave a kiss within the cup
> And I'll not ask for wine.

Upon examining the record more thoroughly, however, sentiments
of human romance go far back into a veritable sea of timelessness.
Ben Johnson's words were a translation of the ancient Greek, which,
in turn, almost certainly was taken from ancient Egyptian love poetry
which expressed these endearments in almost identical terms.

Perhaps we may again appeal to the comparative record for possible
clarification. It is well recognized that birds display before mating
and that squirrels on the lawn enjoy artful chase before the female
ceases, permitting capture. If such be within the realm of romance,
where in the vertebrate lineage might romantic actions first be found?

If the problem of romance be approached in broadest biological perspective, in searching for roots, extra energies for reproduction appear to go as far back as the level of the fish. The question arises
as to what could be considered romantic at this or any other early
evolutionary level. Such actions would need to be sought among the
processes involved in the mating interval, qualifying by virtue of their
added aid to fertilization.

As the piscine reproductive season commences, increased hours of springtime daylight act upon the pineal gland, sending messages for reproductive hormone centers to commence production. The actions of these hormones are twofold. One action stimulates the germinal tissues to produce follicles, eggs in the female and spermatozoa in the male. The other action works upon the body in general, producing elevations of organ activity needed to sustain the steadily increasing work load of follicle maturation.

Germinative work in all vertebrates involves a slowly compounding burden of metabolism. This takes place over a period of time much longer than that of such more transient life functions as digestion, excretion, incidental adaptation or repairing. Preparation for this period of sustained heavy work requires both a full maturation of the system and its maintaining of augmentive work, particularly through increasing its levels of nutrition. Preparation for sexual reproduction eventually comes to include the arranging of the steps necessary for the meeting of the male and female gametes in ways which are suited to the species.

It is characteristic of any biological process involving any essential function that it comes to be endowed with generous margins of reserves. Thus, even when reduced during impoverished times, basic functions are able to proceed. More typically, however, in favorable times they proceed with comfortable working efficiencies. Whether in a fish or other organism, increased levels of reserve are most amply found in the healthiest and most fully mature organisms. Should a courting action be needed, such as through bright coloration or by vigorous body display, the most amply endowed male would be most likely to perform with elegance, becoming thereby the one most likely to charm an attendant female.

A certain potential for romance, so appreciated, appears to be discernable in vertebrates at very early levels. At the level of the fish, the reproductive hormones – primordial estrogen, progesterone and testosterone – act upon a small center in the brainstem, the ventral tegmentum. Activation of this small area, which has been found in human PET scans to light up when subjects have romantic imaginings, sends fresh stimulation to all higher brain levels. Although in fish such levels are but evolutionarily nascent, in more advanced vertebrates such higher stimulation involves the hippocampus with its emotional resources, as well as the forebrain with its cognitive capacities. In such activation the senses are rendered more acute, emotional feelings are mobilized and discriminatory acumen becomes finer. This produces more quickened body movements, while simultaneously the somatic musculature becomes strengthened. These developments take place particularly through actions of testosterone. Adjacent to the ventral tegmentum lies the reticular activating system, through which a wider general mobilization of the entire brain takes place, and, from the brain, of the entire body.

In such a broad view, reproduction, a quintessential life function for every living organism, tends by nature to engage variable and often heavy endowments of energy for the realization of its programmed role in life course progression. Romantic energies appear to occur as an overflow in energies which are already sufficient for mating. Romance, so appreciated, is a dimension of reproductive reserve so expressing itself as to reach such high levels of function as realize natural artistry. High levels of reserve permit functional grace and beauty to arise. Perhaps an image of a displaying peacock or a strutting rooster comes most immediately to mind, or possibly that of a professional dancer.

At the human level, the behavioral reserves in the preliminaries of mating vary extensively, from being entirely absent to involving overwhelming levels of passion. While the roots of human romance appear to be deeply ingrained within the structure of the nervous system, its realization depends upon the presence of favorable circumstances. Civilizations often realize these, but, as American Indian love songs and cultural lore attest, such needs and their gratification have long proven to be experienced as well among hunter-gatherer societies. In Western civilization over the past five hundred years, after the waning of the Dark and Middle Ages, romance has experienced a particular flowering. Among the ancients, Egyptian culture appears to have sustained a long flowering of romantic sentiment. Indeed, a glimpse upon the bust of Queen Nefertiti makes one realize how much the cosmetic industry of today with eye liner, eyebrows, rouge and lipstick, is indebted to this early romantic cultural flowering.

Romantic progression appears to follow two stages of involvement, one earlier in unusually aroused state, and subsequently another, which is more normal. In more intense romantic contexts the body mobilizes stress hormones. These involve adrenal actions at modestly elevated levels. Endorphins are released, which blunt sensibilities, alike to pain and to displeasure. So driven may reproductive processes prove to be in this early stage that, with seeming impunity, endorphins can blunt feelings toward others who are ordinarily close, even those emotionally dear, and can defy rationality itself. At this stage, the levels of the sexual hormones in action remain suboptimal. They act, rather, as if priming for action but are not yet fully involved. In reality, much of romance is often a stress state, as if searching for a sense of interpersonal security or for full assurances of safety before proceeding.

The endocrine profiles of man and woman in romantic love show modest adrenal elevations and suboptimal gonadal hormone levels. Whereas the normal levels of cortisol, the stress hormone, for a man and a woman in a study reported by Maniatis, Well, and Bond were 165 ng/ml and 172 ng/ml respectively, the levels for these persons in early stages of falling in love rose to 224 ng/ml and 243 ng/ml respectively. At the same time, testosterone levels in the male decreased from a normative average of 6.8 ng/ml to 4.1 ng/ml.

This dual phasing of sexual engagement can be noted behaviorally at the level of the reptiles. When anole lizards responding to reproductive scents are approaching one another for mating, the female often halts, remaining poised on the sidelines until the male has cleared the pending copulatory location of potential competitors. Typically she awaits the outcome of a contest in which the loser retreats from the scene and the winner proceeds to mate.

Once intense levels of romantic stress have become resolved or are otherwise absent, the mobilizations of romance advance to states of organization which are highly physiologic. Lovers often note this phase to be more deeply enjoyable. The adrenals participate at lower levels which do not limit but, rather, healthfully advance the full actions of the reproductive hormones. In this range of activity, to the extent to which it becomes realized, the reproductive hormonal complex acts to enliven the entire organism. The senses become quickened, contributing to a more keen awareness. The man enjoys working for the woman, while the woman enjoys the full exercise of her attractive powers before fully committing herself to joining in reproductive action with him.

It is interesting that romantic feelings in humans consistently center about actions of the heart. Even in barest outline, a heart is the symbol of love, in which depiction it has no competitor. The heart is the human symbol for courage, love and anything regarded as spirit. History and the grand lore of conscious experience are too extensive to permit such exclusion to be a last word, no matter what science may say or fail to say. In one sense, it is not surprising that heart awareness tends to occur in romantic experience, since the entire organism at such times is functioning at higher levels of inner mobilization than normal, the increased work of which the heart must produce and then sustain. However, it is the role of the heart normally to function in silence, lest awareness of itself detract from ongoing environmental negotiations.

A physiological explanation may be fairly simple. While the sympathetic nervous division activates cardiac action, parasympathetic nervous action serves to restore the heart between contractions. What appears to take place during mobilizations of love and in strongly affectional states is that both divisions appear to become elevated together, with sensibilities of this increased level of metabolic activity being reported to consciousness as remaining physiologically healthful. Involving no pathology, it is not surprising that such states have failed so far to become medically studied.

Although unlikely to be found at the center of awareness, in a longer range perspective the surge of energies which attend romantic thoughts ultimately direct to parenting. Commonly such thoughts lie at first in the background. However, often one partner chooses a particular other for precisely this reason, sensing that he or she would be an auspicious candidate for bearing and raising children. When romance flourishes, its sentiments orient one person to a particular other with a shared sense of exclusiveness. This, in time, will lead to parenting. The enhanced energies which are released in romantic states serve to overcome obstacles, including some which may even be quite personal. Of the faults of the special other, love becomes proverbially blind. If such do exist, certainly the future will take care of them.

If we of today would desire to see the human future, we would not go wrong in attempting to envision how romantic nature might further evolve. This would take place not as an experience of wild, reckless, irrational energy but in enhanced realizations of greater inner resources working to favor increased fulfillment in bonding.

What might be the characteristics of an enhanced role of romantic forces in daily human experience? Although their specification, once identified, may seem so simple as but to portray common sense, their features are not always immediately obvious beforehand.

Furthermore, an aura of the inexplicable often attends romance, as it does in attempting to understand choices in mating. Large unconscious forces are at work.

Sensing that the human male is primarily a visual creature, not too far removed from a more reflexive animal state, a woman who desires romance finds pleasure in making and keeping herself physically attractive. She likes to be clean and to dress appealingly. When appropriate, she may also learn to undress herself in stages which artfully appeal. She inherently senses that slovenliness and lack of cleanliness are unromantic.

More essential, however, is her personal role toward a particular man. She needs to be open to his needs and to respond to his individual nature with interest. A shared sense of humor acts as a welcome bond, as its lack tends to prove a drain. She knows how to use the sense of touch for its special powers of appreciation and attraction. Perhaps most important of all is a sense of personal appreciation and encouragement. No greater compliment occurs within a man's psyche than for a woman of his affections to feel him to be wonderful.

Romantic impulses in the man orient toward pleasing his woman. The exercise of even strenuous energies for work, any tasks undertaken in her behalf become inherently pleasing. A romantically attuned woman senses such efforts and responds to them appreciatively. Failure to appreciate them tends to dampen male desire. On the other hand, excessive appreciation in the form of flattery often detracts, particularly if unmasked.

In romance, there is a role for each to play. The man's role is to let woman know that he finds her lovely and beautiful, first in attractiveness but yet more as a person. She motivates him. As romantic nature involves extra energies for mating, such energies commonly involve unusual willingness for effort and the most delicate sensibilities of which the person is capable. Though they may arise unexpectedly, romantic impulses cannot be taken for granted. For them to last, nurturance on the part of each is necessary. Without mutuality their freshness wanes and their very essences disappear. It is woman's role to sense the game and to further it as desired. Over time, broad personality development, with abilities not to let interrelatonal problems fester and with well developed capacities for forgiveness help immeasurably.

It is interesting, nevertheless, how effectively romantic thoughts and impulses are able to persist, enlivening daily events throughout the lifetime. Reflecting this, the enormous profusion of women's magazines presents a self-explanatory aspect of the early years of maturity. The vast lines of romance novels in the library continue to circulate most widely of all. Even in elderly women in nursing homes, the reading shelves consist mainly of romance novels. When Queen Kapiolani's husband left Hawaii to seek medical attention in the continental United States, though well along in years, she wrote love poems, setting them to such hauntingly beautiful music composed by herself, that they are still popular today. Similarly, many a man well along in years refuses to be inwardly changed, preferring to remain in his innermost self an unreconstructed romantic, preferring, he will say, not to grow old.

Romance, so appreciated, is well-ingrained reproductive nature expressing itself as art in gender relationships. The world of romance is one of need elevated to levels of sustained desire which, when shared, find pleasure in mutual attention and service. Romance engenders a sense of a more fulfilled reproductive life. A child of romance often comes as a special gift.

It is quite possible that civilization, as feared, could descend into new dark ages. If so, unless vertebrates themselves should disappear, it is difficult to envision that capacities for romance would entirely disappear. Rather, with roots so deep in vertebrate nature, such basic capacities would seem likely to remain for future realization. This at human levels requires alike, favorable circumstance, well-attuned interpersonal discrimination and a lasting desire to keep gender relationships alive. Though romance is beneficial to humans establishing well-bonded states, it remains helpful experienced at lesser intensities throughout the lifetime.

In a world of such vast imperfections as that which we have inherited, we must be thankful for whatever of shared romance our individual lives experience, the memories of which crown an aged brow and uniquely endure to the end.

XVII

THE FOURTH HUMAN FUNCTIONAL DOMAIN

The ultimate result of sexual privacy is the creation of a loving human household.

*I*t has been supported that the advent of privacy as a prerequisite to the human sexual relationship served to provide access to a new realm of interpersonal dynamics between man and woman, such as had never existed before. If distinctly human reproduction commenced with sexual privacy, what, then, has happened to this uniquely human dimension in the subsequent two million years?

The emergent inner fourth domain of human experience presents a world unique unto itself. Whereas the distal domain, the proximal domain and that of the family at home base are each spatially defined, that of the fourth domain is almost entirely defined by function. In the few instances where the fourth domain approaches spatial definition it occurs in such polygamous societies as the rainforest pygmies and the African Bantu where each wife is given her special hut in the area of home base. Clearly appreciating the private nature of the inner fourth domain, when they were designing symbols for a written language the ancient Chinese chose for happiness a representation of one woman under one roof.

In most societies over the world today the fourth domain, variously developed, remains guardedly private. Inwardly experiencing this domain, every well-married person in Western culture moves about daily, preserving its essence, unseen and unspoken to others. Like the sexual relationship from which it takes its origin, a couple's realm of interpersonal development is hoarded, at once too prized to share with others, lest perhaps someone steal it, and too personal to be broadcast. Perhaps the most evident outward sign of its presence is the measure of comfort and gentle ease with which a man and woman speak to each other with implicit understanding.

The lengthening of the duration of sexual intercourse led to realizations of sexual love. The emotions elicited during intromission prove to be as powerful as any in human nature. In sexual loving, each, in appreciation, desires to give to the other. Through the personal interactions between a man and a woman as they experience rhythms of engagement and pleasures together, a sense of shared identity progressively emerges, a mutually experienced inner world. In broad behavioral terms, this comprises a humanly emergent fourth realm of experience, functioning at the center of the home domain. At a practical level, this involves care for each other, shared concern with the development of mutual progeny, and engagement in the wider functions of the family unit, of which they serve as the heads.

With intimacy, as noted, the other person, one's partner comes to be part of a larger self. This is foremost an emotional identification and only secondarily rational. Each comes to live both in an individual realm of experience and in a shared realm of experience with the other. The development of emotional identity is reinforced by sexual continuity. Although the emotions aroused during intercourse are felt to be immediately all encompassing, the development of a sense of shared identity progresses by degree, by steps and stages over time.

An interesting experiment relating to the capacity of sexual intercourse to produce intimacy in favorable circumstances was reported recently by a couple about to enter the threshold to middle age. The man's age, 40 years, was at a position in life not usually associated with improved sexual bonding. The marriage had been sound, with a hearty sexual life and favorable emotional compatibility. However, the wife had noted a steady decline of sexual interest on the part of her husband. Reflecting happily upon their past good sexual years, she decided to offer him as a birthday gift daily sex for 100 days, with the sole proviso that there would be no exceptions. No hemming and hawing, no urgent preoccupations, no abstentions.

By the time that this date arrived, each had become to the other a person whom neither partner had ever before sensed or anticipated. As their sexual natures seemed to deepen, a new level of the relationship came to be realized. A daily forgiveness and new kindness were reported, with a total absence of irritability. Annoyances faded away. Each became highly aware of the presence or absence of the other. Each seemed to have a sixth sense of where the other was - emotionally, physically, and mentally.

After the 100 day goal had been reached, nightly sex proved to be too demanding to be continued. The couple reverted to approximately two days a week, a level more comfortably appropriate to their ages. However, upon continuing at this less intense level the couple noted an improvement in the quality of their sexual life. It is as if, in physiological terms, elevated levels of oxytocin from more frequent and prolonged sexual involvement had elevated their personalities to higher levels of daily life management.

In the context of current Western culture, one is inclined to regard this experiment, perhaps, as an interesting curiosity. In the context of the present study, however, this experiment appears to reveal that in appropriate couples deeper levels of interpersonal fusion become possible with a deepening of sexual relations even in sexually declining years, an inheritance from our ancestral hunter-gatherer past.

A deeply ingrained theme of human reproductive nature relates to departure and return. When the sense of loving becomes sufficiently strong, a sense of the presence of the other may develop when they are apart. Each seems to be aware of where the other is and what that person is doing. When it develops, such a sense derives from integrated feelings of loving, of spatial distance, and from feelings of bonded continuity. This is perhaps most intensely experienced in the early years of a paired relationship. It may occur, however, with any significant separation, perhaps most commonly during the years when a woman is most heavily dependent upon a man for the care of their young. However, where forces of loving are strong, it can occur at any age.

On such occasions, as when the hunter is far from home, he may feel as if his beloved were at his side, arm about his waist, supporting his efforts, at once encouraging him on and protecting him in the face of danger. While in naturalistic terms this is felt as a sense of presence, in terms of the hunter-gatherer world-view it is felt as the spirit of his beloved accompanying him on the chase.

Setting aside her anxieties, woman at home base remains aware of the risks and dangers of the unpredictable outer world. Separation serves to engender an appreciative welcome upon his return, most effectively expressed by warm personal reaffirmation in privacy.

It is an interesting feature of our hunter-gatherer heritage that, in spite of risks and dangers, as part of his daily routine the human male feels impelled to leave home. The evolution of hunter-gather nature has essentially mandated this by rendering him to be immediately refractory after sexual intercourse, a distinctly human male reproductive development. No comparable state evolved in the human female or appears to exist at simian levels. Were he not in a sexually refractory state, the human male might not ever want to leave home. Here evolving human nature worked both by bait and switch. Thus, should he remain at home, there would be no special rejoicing upon his return, no proud but modest bearing of game to be shared, no special tale for evening narration, and, though it would be unlikely to be refused, sex would not have its full luster. The primal human hero is man returning home successful in the hunt.

Complementary to this, the human male, particularly the youthful male, experiences a compelling need to master the external world to levels sufficient to maintain a family. The ancient Zulus would not allow a man to marry until, single handedly, he had killed a lion. In today's world, fascination with sports and souped up cars are expressions of this deep inner need for external mastery.

A woman's sense of wider adventure in the hunter-gatherer world finds its main enjoyment in a man's distal environmental experiences. For her to remain at home much of the day is usually not an uncomfortable task. Her venturing away from home is far less necessary. However, she does need to forage for food almost daily in areas near to home base. It is also an essential part of her nature that she should maintain an ongoing awareness of the larger outer realm as well. Her foraging excursions often discover valuable information for the hunters, particularly the locations of nearby game and their directions of movement. It is also important for her to have a working knowledge of the resources of the wide terrain, such as the locations of rare sources of water or of nuts still on trees in periods of drought. A basic knowledge of the male's outer realm helps man and woman to communicate more widely, with deeper appreciation of each other's roles.

Although modern culture has modified the details of primal couple separation almost beyond recognition, certain predispositions toward departure and return timelessly persist. In sentiments desiring sexual exclusiveness the popular songstress, Taylor Swift, anticipating her man's return, croons, "Please don't be in love with anyone else. Please don't have anyone else waiting on you".

Departure and return find their reward at journey's end in renewed sexual bonding, a state in which intercourse has advanced to serve as a basic sustaining activity. The essence of home for humans is no longer a mere site for safety, or a return to the interpersonally busy life of a band, but where man and woman may live as a paired family unit in close harmony.

Over the years, experiences of loving come to know many moods. Particularly for an older male, an endearing aspect of home return is that a well motivated and well accustomed partner often has power to rejuvenate him, thus to complete the day with sexual reward. In hunter-gatherer contexts this implicitly enhances his sense of manliness. In ecologic perspective, he becomes thereby better prepared for meeting familial needs the next day. In personal perspective, his life becomes more fulfilled.

Perhaps modern life has lost cultural awareness of this ancient human conjugal theme As the hunter-gatherer way of life first merged into village life in the ancient Nile valley, and thence into a first civilization, no good Egyptian woman facing eternity with a sculptured representation of herself would want to be presented in any way other than with her arm supportively placed about her husband's waist. Though inherited from its hunter-gatherer prehistoric past, this theme persisted throughout all of ancient Egyptian history, over two thousand years.

Physiologists describe the human organism as a comfort machine. The living system is seen to rest in homeostatic balance. An imbalance arises, which comes to be sensed as a need. The imbalanced system finds that it is endowed with resources to correct the imbalance. The need is met, following which comfort and balance are restored.

This view fails to appreciate that from the moment of waking in the morning the living system is characterized by an overflow of autonomous energies, lacking which one is ill or dead. One example of such an abundance of activation energies occurs in those which impel him to leave home. A related example of such energetic overflow lies in his enjoyment of the challenge of the long endured chase and in the exercise of his autonomous capabilities. In all fairness, it must be appreciated that physiology also explores various states of activation. However, there is much to commend physiology's basic position. What commenced in human evolution as a transient location for women and children in a nomadic way of life, a home base, has advanced at the fully human level to become a center of all that is finally satisfying in life.

What is home? Historically, the emergence and development of the fourth ecologic domain first created and then steadily enlarged the meaning of home. Home is a dwelling, indeed, but in a more generic sense home is the domain of the inner vital processes that most advance life. To the advancement of these, all external activities are ultimately subordinate. Human physiology basically divides between those actions relating to the outer world and those of the inner. The former comprise the somatic, voluntary division, the latter the visceral, autonomous division. Home speaks for a functional summation of all antecedent activity, resulting in serial fulfillments of the visceral, vital self, each action of which serves to perpetuate or to advance life by an integral step.

Home is where the ancient continuity of life becomes most experienced. Here in contexts of family each person becomes able to sense the grand human procession of which every one is a part.

In one sense, the position of the physiologist is unassailable. Comfort is the representation in awareness of favorable inner stability, which, in turn, signifies optimal organic continuity. Organic continuity, now over 3.8 billion years in existence on the planet Earth and prospering well, is a central description of life itself.

XVIII

HUMAN NATURE'S
FOUR ECOLOGIC REALMS

Hunter-gatherer nature developed to function alike in
the wider macrocosm, in the proximal microcosm, in
the home base area, and at the special center of the
human family.

*W*hereas the ecologic domain of a great ape involves one functional world in which the entire band moves together, as noted, the
functional ecology of humans has advanced and diversified over the
past two million years to involve four realms. These range from the
most spatially distal to the most closely intimate. Each involves a different work apportionment for the general advancement of the group
or the species. Their relatively simple summation at modern levels,
reflecting their human evolutionary past, may be briefly noted.

The distal outer environment is primarily man's realm. In hunter-gatherer society it often covered hundreds of square miles. Here food and
raw materials were obtained, and band safety was secured. In attaining these, such masculine attributes as spatial discrimination, large
motor skills and physical endurance evolved, primarily in males.
Emotional elaboration centered upon such major exertions as take
place in acute excitement, hostility and fear. Some mutual bonding

takes place among males who hunt or otherwise work together. Woman's engagement with this realm is minor, while that of children and infants is prohibited.

The proximal environment about home base served primarily as woman's domain. Here women move together daily, gathering vegetable foods, firewood, water and other domestic essentials. Almost constant vocalization takes place among the gatherers, as well as with older children who are present. While infants are carried in skins for convenience of nursing, other children are left at home bases. Men know this realm, but are only secondarily involved in its use.

Home base serves as the domain of the entire family. Here basically the children are raised in safety. Mothers teach them language and culture. Women tend the fire, cook, nurture the young and care for the indisposed. Home base is the grand functional center for the activities and satisfactions of all family members, young and old.

A fourth human domain, defined not by spatial reference but, rather, entirely by function, exists at the center of the family. This realm is formed by the parental bond, to the exclusion of all else. Within the family itself, incest prohibitions serve to maintain the exclusiveness of this domain. Without the satisfactory function of the fourth domain, that of the sexual privacy and intimate meanings which parent personalities together develop, human function in all other domains proves to be less fully defined and suboptimally effective.

More than any other feature of human evolution, developments in the fourth domain are predicated upon advances in privacy. As the existence of privacy makes possible fully human experience within the fourth domain, continuing privacy makes possible its duration. With loss of privacy the fourth domain disappears and sexual expression ceases to be fully human.

PART V: GENERAL CONSIDERATIONS

XIX

THE SEXUAL MICROCOSM AND THE LARGER MACROCOSM

The private sexual domain remains ever but a portion of the larger realm of the broad self-maintaining activities of the culture.

*T*he personal world of sexual privacy and the wider external realm exist in figure- ground relationship. Each is a domain partly influenced by the other. In any final analysis, the sexual realm is only a portion of the much larger world of life experience, which readily enhances, may diminish, or quite eclipse it. This is a subject too enormous to be dwelt upon here, but one that needs appropriate recognition, for the human sexual realm, no matter how developed, never exists entirely alone.

In a harsh or challenging environment, the rewards of hearth and home serve as basic compensations, affirming that life is good and renewing capacities to perform difficult environmental tasks. In many ways, the toils and strains of today's living seem not unlike those of human Ice Age experience, in which comforts and rewards of home are sought in compensation for the heavy exactions of everyday life. On the other hand, in a context of abundance and generosity, the richness of the private inner realm tends to overflow into outer activities, adding grace and having the power to render the mundane special.

On the whole, as a first statement of relation, the world of sexual privacy works to contribute favorably to society. Within the family, needs for privacy isolate the sexual. A well developed inner life of couples serves also to ease any of that pair's frictions with the larger group. The societal consequences of well-functioning sexual privacy are widespread. At a crucial moment in a hunt, perhaps while waiting for a seal to rise for air through a hole made in the ice, or in the company of a band hunting an elephant from downwind, the slightest interpersonal friction expressed unknowingly or impatiently at an inopportune moment can prove disastrous for an outcome. For women gathering foods together, it is best for one who is sexually available not to elicit envy or anxiety in others. In all human cultures it is best for all sexual couples, indeed, for all adults to be as well bonded sexually as possible. As the psychologists report, in whatever may be the culture, a sexually satisfied person is different from one unsatisfied. A basic measure of any culture is how well its couples become bonded.

Viewed alternately, for a sexual relationship to last, even one with a highly favorable emotional composition, it proves to be essential that the individual roles of the pair fit satisfactorily within the broader culture. Even more favorably, sexual relationships fare the best when its individuals thrive within the culture.

XX

PROBLEMS INTRODUCED BY SEXUAL PRIVACY

> With its enormous complexity, sexual privacy acts as
> a double edged sword. When things are going well,
> sexual privacy liberates human experience to enter
> a new world of wonders. However, when things are
> in trouble, privacy's encapsulation renders problems
> more difficult to resolve.

*S*exual arousal is a special state. The importance of sexual experience for the continuity of the species is indicated by the enormity of the emotional powers and the generous physiological resources at its call. When the course of a sexual experience goes well, it can proceed artlessly, with a refreshing loveliness comparable to an idyllic summer day. In such situations the complexity of the underlying physiological organization making it possible is quite inapparent; all seems to be so simple. Let problems arise, however, and they may unfold as troubles without end. If, as the ancients believed, the gods played sport with humans, the major arena for this would appear to lie in human sexual nature.

As if aware of such vulnerabilities, human evolution has developed certain protections. During acute arousal endorphins are produced, the body's indigenous morphine-like system, through which acute pain is blunted and modest pain temporarily obliterated. As sexual learning progresses along with personal maturation, improvements in rationality help one to realize that problems will arise and, perceiving them in a enlarging perspective, one becomes better able to outgrow them. Time also helps with lesser problems, tending to forget them, particularly when they exist within contexts of well-fulfilled experience.

A broad perspective upon human sexual problems might view the general landscape as follows. Humans are by nature sexual beings whose life course for each person involves sexual learning. Sexual learning at any point is mainly guided by innate feelings of pleasure. By producing endorphans, the state of arousal protects against certain levels of unpleasntness, thereby assisting the reproductive action in progress to proceed to its completion. Thus in the aroused state of intercourse individuals may perform acts which under ordinary circumstances they would not find themselves doing. It is the sexually more traumatic and more interpersonally injurious experiences that tend to present unresolved problems. These problems also are always somewhat special, in that they tend to involve not one but two individuals, in deeply personal and emotional ways. They tend to be endowed with unusual potency.

It is not easy for individuals to work out sexual problems of larger proportions between themselves. Some of these problems can be so strong as to modify the entire personality, at times for life. A rejected woman's countenance may entirely change. Though walled from awareness for much or most of the time, unresolved sexual problems do not remain static. Rather, they continue to act within the personality. A common consequence is for the experience of uninhibited pleasure to become reduced. Not only in regard to sex, but one's partner cannot be enjoyed as much. If problems are sufficiently severe, a full sexual burnout may result. The worst instances are those of sexual violence experienced by the young. Many a young person never completely gets over such trauma.

Even when visiting a doctor in his office, one does not ordinarily discuss sexual problems unless they are specifically brought up, and even then a few words are usually expected to say a great deal. In general, even though dress and the entertainment industry may proclaim it to be otherwise, a sexual reticence pertains for the culture. Only when one notes a tattoo, perhaps, on a woman's shoulder or leg may we consciously appreciate that, here too, in willing expression is a sexual being.

Without attempting to delve too far into the therapist's realm, it is possible to appreciate certain basic steps which help to resolve problems involving sexual conflict. To be effective, the approach needs to be one sufficiently well motivated and desiring resolution in a spirit of open-mindedness. Scolding at such a point serves only to close the mind and to terminate any attempt to resolve issues. The basic facts need to be elicited and placed on the table as objectively as possible. These are best understood from the shared perspectives of both individuals, each of whom tends to see events differently. Sometimes, however, merely a deeper understanding may be sufficient to resolve issues. Many a bunny comes to feel in time that she is being used, and with such realization becomes wiser in regard to pursuing a more sensible future.

There arises the need to ascertain where the crux of a problem lies. This step may particularly benefit from an external point of view, providing an outside sense of proportion and of appropriateness. Attempting to find forgiveness, a subsequent stage, can prove to be extremely difficult. Only as one person forgives the other does full forgiveness develop, with that of the self often proving to be the hardest. The entire process is a tall order. It has sent many a person to a counselor, to a confessional or to turn to deep prayer. Although mutual pardons are not always possible or even to be expected, what is most desirable is that each supplicant feel that he or she has done whatever may be possible to understand, including to make amends and to change course in the future. It is important that the supplicant fully sense and acknowledge his or her responsibility, which in itself can prove to be sufficient to bring about resolution.

These are not simple or easy problems. Whenever possible, they are best handled at the time when they occur. Neither as individuals nor as a culture do we appear to have the facility for meeting these problems which we should have. Though it is not clear how, we need somehow to do better.

In the long range, the most effective course remains ever the prevention of injury. For this, understanding helps. First sexual experiences tend to be highly formative. In order to avoid emotional or physical trauma in a first sexual experience, traditional African culture worked out certain solutions. Before marriage, a young woman is visited by an older married cousin, a female of the same generation but one not a member of the same household. With her own experience to guide her, she explains to the prospective bride a man's intense sexual nature at this age, driven and with little patience. She is advised concerning how to anticipate it and how best to meet it. Meanwhile, the prospective groom is visited by an older male who will explain to him woman's slower nature of arousal, with its need for male self-control and gentle patience. Male ardency, an understanding senior might advise each, is nature's method of establishing a family.

XXI

ENHANCED CARE OF PROGENY AS A MAJOR VERTEBRATE TREND

The overall trend of reproductive advances in vertebrate evolution appears to direct with marked consistency toward increased biological investment in the
unit progeny of the species.

*C*an the various strands of interpretation that have been followed in
this study be gathered into any salient features?

It is commonly accepted that in the long evolutional progression from
the level of the amphibian to the human the crucial reproductive
advance was the emergence of internal fertilization. It may perhaps
be somewhat humbling for a modern human, whose adult life is so
much involved with sexual intercourse as a recurrent maintenance
activity, to have to accept that this assumption appears to be not quite
correct. The major advance in the reproductive trajectory of higher
vertebrates appears to have been steady biological investment in the
unit progeny of a species, in relation to which internal fertilization
secondary evolved.

In the interpretation of this study, the source of the intense, pervasive,
and often obscure human need for sexual privacy ultimately was the

191 ►

need to render increased biological investment in the developing human young, and, as a result, the development of more adept human adults.

It is impressive and perhaps surprising how evolutionally ancient and remarkably consistent this general trend reveals itself to be. It may be noted first in the emergence of early vertebrates as they migrated onto land from the seas. As predatory pressures in land environments proved to be less, and, stage by stage, as organisms enjoyed improved land adeptness, increased biological investment in the unit progeny of a species became possible. If generous allowances may be made for the existence of wide variation in egg counts that can be here only approximated, in its general landmarks this trend may be readily noted. For species maintenance in the ocean, the pelagic fish must broadcast million-fold bare eggs into the surrounding waters. The freshwater fish, in comparison, achieves species stability spawning mere thousand-fold eggs, each somewhat larger and usually able to be slightly endowed with nutrition. The amphibian, in turn, lays its clusters of hundred-fold eggs, each much larger and surrounded by a generous coat of protective jelly. The reptile lays its dramatically advanced amniote eggs by the dozens, each heavily endowed internally with nutrition and externally protected by a shell. At the level of an early mammal, a mere three to six offspring per gestation becomes possible, with each infant nursed after birth for one to three months. At the level of the great ape, one infant is born after eight months of gestation, to be nursed for three to four years and parented from five to eight. These generalized comparative landmarks are but the surface aspects of myriad reproductive advances which have been taking place at multiple functional levels in many vertebrate lines over the last 450 million years. Each major evolutionary progression in land adaptation has been characterized by a significant increase in the biological endowment of the unit progeny of the participant species.

Human evolution appears to have advanced this ancient trend further.

The biological investment in each human offspring has become so advanced that often all members of our species are argued to be engaged in reproduction all of the time. Such an assertion takes a position to its limit, and yet appears to be not without certain validity. After nine months of gestation and a period of nursing which may extend to three years or more, the care of each human offspring continues through two decades. Yet even here it does not cease. Studies reveal that even attentions from grandparents tend to lengthen the life spans of the maturing young.

It is interesting that, although its progression has not advanced anywhere nearly as far, the evolution of plant life on land reveals trends similar to those of animal evolution. In brief, a fern, a primitive plant of the water's edge, for maintaining its species perpetuation broadcasts its miniscule spores by the many millions. In comparison, an oak, a land plant physiologically advanced to withstand hurricane, drought and winter frost, maintains species stability producing acorns by the mere thousands. This presents an enormous advance in the biological proficiency of the unit seed. If the fern reveals similarities to the reproductive level of the fish, in many ways the acorn of the oak resembles the reptilian amniote egg. A strong outer coat or shell, endowed within with a rich supply of yolk and protein for the seedling to use for its initial nourishment up to and immediately after sprouting. In addition to the dramatic advance which produced the amniote egg, one is reminded of certain levels of aftercare in some reptiles. After hearing their young as they hatch, crocodile parents carry their hatchlings in their mouth gently from dangerous and nutritionally barren sandy nest locations to rushes, where each may safely grow amidst an ample availability of food. Such behavior is functionally equivalent to the early nourishment of the oak seedling that is provided by the cotyledon on the stalk.

As both animal and plant evolution demonstrate, increased biological investment in the unit offspring of a species does not occur as solitary change as much as one component among steps following more general biological trends. Often changes tend to occur as syndromes. The major interactive components of evolutionary change include the availability of food, the relative proficiency of an organism's adaptive accommodation to its environment, and reproductive losses due to predation. In the above profiles, improved adaptive proficiency proves likely to lead to nutritional improvement, as nutritional improvement aids adaptation. Improvements in one or both of these spheres, in turn, tend to lead to reproductive improvements. Especially where increased numbers of progeny are no longer needed for species maintenance, qualitative improvements in unit progeny become possible. These, in turn, tend to produce improved capacities for adaptation. In such simplified terms, all major advances tend to evolve in broad interaction.

It is now well accepted that evolution takes place through differential reproduction. Due to variation among the offspring of a species, some individuals will adapt and prosper within a given environment better than others.

From such persistent trends in the vertebrate past can one anticipate the human reproductive future? Having attained total land conquest, have humans, perhaps, come to a reproductive terminus? Will human evolution somehow continue, or do present conditions portend some kind of transition?

It is ever difficult, if not hazardous, to attempt to predict the future from the past, particularly when its variables have been complex and it may be approaching a newly defining stage. However, this study suggests certain major evolutionary trends which appear likely to continue.

It would appear more fitting to assume that the human lineage in evolution, which has been accelerating in the past six million years, will somehow continue to elaborate. In particular, two trends of major relevance appear to be the increasing biological investment in unit progeny with external environmental advancement and the well-recognized ecological trend toward achieving species stability after each significant advance.

The grand evolutional trend toward increased investment in the unit progeny of a higher vertebrate species in conjunction with the recurring ecologic trend toward improved species stability appears to project toward an end point approximating two children per couple. Somewhat more than two might be needed to compensate for childhood losses, which at such points could prove to be relatively few. Slightly more than two might also be needed to compensate for couples who are infertile or otherwise fail to reproduce. Comprising an essential familial foursome, such a pattern subsumes the attaining of a highly favorable human ecologic balance.

Perhaps the fully human familial plan toward which vertebrate evolution is moving would consist of two parents and two children. In a highly favorable ecologic balance, such a familial pattern, based upon more advanced and fulfilling parental roles, promises optimal attention from each parent to each child. These considerations suggest potentials for domestic felicity on a broader and deeper scale than that which has generally been realized – greater, perhaps, than any that has yet prevailed in human history.

SELECTED REFERENCES

1. Bayley, N. Growth curves by height and weight by age for boys and girls, scaled according to physical maturity. Journal of Pediatrics 1956; 48: 187-194.
2. Bem, S. The measurement of psychological androgeny. Journal of Consulting and Clinical Psychology 1974; 42(2): 155-162.
3. Brizendine, L. The Female Brain. Broadway Books, New York, New York. 2006.
4. DeWaal, F. Bonobo, The Forgotten Ape. University of California Press, Berkeley, California. 1997.
5. Diminis, J. and Eddy, M. The Cats of Africa. Time-Life Books, New York, New York. 1963.
6. Donaldson, F. Emotion as an accessory vital system. Perspectives in Biology and Medicine 1971;15: 48-71.
7. Dybas, C. Leopards in the twilight zone. Natural History 2011; 119: 30-37.
8. Eldredge, N. Life Pulse. Facts on File, New York. 1987.
9. Ellison, P. (Ed.) Reproductive Ecology and Human Evolution. New York, Aldine de Gruyter. 2001.
10. Gibbons, A. The First Human: The Race to discover our Earliest Ancestors. New York, Anchor Books.2006.
11. Goodall, J. In the Shadow of Man. (Rev. Ed.) Boston, Houghton Mifflin, 1983.
12. Goodall, J. The Chimpanzees of Gombe. Belknap, Harvard University Press, Cambridge, Massachusetts. 1986.

13. Guyton, A. Textbook of Medical Physiology, 6th Ed. Philadelphia, Saunders.1983.

14. Hart, D. and Sussman, R. Man the Hunted: Primates, Predators, and Human Evolution. Westview Press, New York, New York. 2005.

15. Hrdy, S. Mothers and Others. Belknap, Harvard University Press, Cambridge, Massachusetts. 2009.

16. James, W. Principles of Psychology, Volume 2. Holt, New York, New York.1890.

17. Jolly, A. The Evolution of Primate Behavior. Macmillan, New York, New York. 1977.

18. Komisaruk, B., Beyer-Flores, C. and Whipple, B. The Science of Orgasm. Johns Hopkins Press, Baltimore, Maryland. 2006.

19. Johanson, D. and Wong, K. Lucy's Legacy. Harmony, New York, New York. 2009.

20. Kinsey, A., Pomeroy, W., Martin, C. Sexual Behavior in the Human Male. University of Indiana Press, Bloomington, Indiana. 1948.

21. Kinsey, A., Pomeroy, W., et al. Sexual Behavior in the Human Female. Philadelphia, Saunders. 1953.

22. Langmuir, and Broecker, W. How to Build a Habitable Planet. (Rev. Ed.) University of Chicago Press, Chicago, Illinois. 2012.

23. Leakey, R. and Lewin, R. Origins. Macdonald and James, London. 1977.

24. Lee, R. and DeVore, I. [Eds.] Man the Hunter. Aldine, New York. 1969.

25. Lester, D. Comparative Psychology. Alfred Publishing Company, Port Washington, New York. 1973.

26. Lieberman, D. The Evolution of the Human Head. Belknap, Harvard University Press, Cambridge, Massachusetts. 2011.

27. Lovejoy, O., Suwa, G., Simpson, S., et al. The great divide: *Ardipithecus ramidus* reveals the postcrania of our last common ancestor with the African apes. Science 2009; 65236: 100-106.

28. MacLean, P. New findings relevant to the evolution of

psychosexual functions of the brain. Journal of Nervous and Mental Disease, 1963: 135: 4, 289-301.

29. McGlone, F., Valibo, A., et al. Discriminative touch and emotional touch. Canadian Journal of Experimental Psychology 2007; 61 (3): 173-183.

30. Montagu, A. The Direction of Human Development, Rev. Ed. Hawthorne, New York. 1970.

31. Naessen, T., Lindmark, B.,Lagerstrom, C., et al. Early postmeno-pausal hormone therapy improves postural balance. Menopause 2007; 14 (1): 14-19.

32. Napier, J., and P. Napier. The Natural History of the Primates. MIT Press, Cambridge, Massachusetts.1985.

33. Newman, R. Why man is such a sweaty and thirsty naked animal: A speculative review. In, Montagu, A.Ed), The Origin and Evolution of Man. Thomas Crowell, New York, New York. 1973.

34. Papez, J. A proposed mechanism of emotion. AMA Archives of Neurology and Psychiatry; 1937: 38, 725-743.

35. Roberts, A. Evolution: The Human Story. DK Publishing, New York, New York.2011.

36. Russell, R. The Lemur's Legacy. Putnam, New York, New York. 1991.

37. Shostak, M. Nisa, The Life and Words of a !Kung Woman. Vantage Press, Random House, New York, New York. 1983.

38. Stebins, R. and Cohen, N. A Natural History of Amphibians. Princeton University Press, Princeton, New Jersey. 1995.

39. Smith, H. From Fish to Philosopher. Little Brown, Boston, Massachusetts. 1953

40. Stringer, C., Lone Survivors: How We Came to be the Only Humans on Earth. New York, New York. New York Times Books, Henry Holt. 2012.

41. Tattersall, I. Masters of the Planet: The Search for our Human Origins. New York, New York. Palgrave Macmillan. 2012.

42. Terborgh, J. Diversity and the Tropical Rain Forest. Scientific American Library. New York, New York. 1992.

43. Thomas, E. The Old Way: A Study of the First People. Farrar, Straus, and Giroux. New York, New York. 2006.

44. Van Der Post, L. and Taylor, L. Testament to the Bushmen. Penguin, New York, New York. 1985.

45. Wald, G. 1954. The origin of life. Scientific American, 191: 45-53.

46. Wells, S. Pandora's Box: The Unforeseen Costs of Civilization. Random House, New York, New York. 2010.

47. White, T., Asfaw, B., Beyone, Y. et al. Ardipithecus and the paleobiology of early hominids. Science 2009; 326: 75-86.

48. Whitehead, A. Modes of Thought. Free Press, Macmillan, Toronto. 1968.

49. Wong, K. First of our kind. Scientific American 2012; 306: 30-39.

50. Wrangham, R. Catching Fire: How Cooking Made Us Human. Basic Books, New York, New York. 2009.

PART VI: SYNOPSIS
THE EVOLUTION OF HUMAN
SEXUAL PRIVACY

SYNOPSIS: THE EVOLUTION OF HUMAN SEXUAL PRIVACY

The human paleontologic record attained through the study of stones and bones is now sufficiently advanced to permit early interpretation of related prehuman and human soft tissue evolutionary developments. Focusing upon reproduction, this study supports that a critical turning point occurred with the advent of sexual privacy.

Before privacy emerged at the beginning of the human career, new needs for safety, emotional involvement, and more specialized uses of touch had emerged at earlier hominid levels. As human nature commenced to become more environmentally adept after the Ice Ages descended, it became urgent for fathers increasingly to give steady, conscientious care to their progeny. The years needed for development of human young during the human tenure expanded from approximately eight to twenty.

The incorporation of the father into the family unit was brought about mainly by human reproductive advances. In this novel program, the emergence of sexual privacy opened up to human nature a new inner universe of experience and meaning. The sexual experience was permitted to have potentially unlimited duration, and became intensified by orgasms. Associated emotional features advanced from being merely appreciative to becoming loving and enduringly affectionate. The cognitive aspects of the sexual experience advanced to become potentially the most intimately personal of the human lifetime. As the years needed for the maturation of human young increased, needs for privacy intensified, buttressing parental stability and fostering longer range constancy.

Many features of human reproductive nature that are obscure in present day culture find explanation in our hunter-gatherer past. Woman's slowness in sexual arousal, man's sexual drivenness, the meaning of the female orgasm, problems of sexual constancy, gender differences in sexual satisfaction, the male's need for environmental mastery, the depth of human familial nature – all such modern reproductive features whose nature is not always self- evident find explanation in our aboriginal human past.

It is not possible to return to the ancient hunter-gatherer world within which our human reproductive natures arose, but perhaps a better understanding of their origin may enable us to see our way forward to a more fulfilling future.